DATE DUE

Drinking Water Hazards

How to Know if There Are Toxic Chemicals in Your Water and What to Do if There Are

By John Cary Stewart

ENVIROGRAPHICS

Hiram, Ohio

Drinking Water Hazards

How to Know if There Are Toxic Chemicals in Your Water and What to Do if There Are

By John Cary Stewart

Published by:
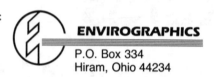
ENVIROGRAPHICS

P.O. Box 334
Hiram, Ohio 44234

Printed in the United States of America on recycled paper.

Library of Congress Cataloging-in-Publication Data
Stewart, John Cary
    Drinking water hazards.

    Includes index.
    1. Drinking water — Contamination.   2. Drinking water — Analysis.
    3. Drinking water — Purification.   4. Consumer education.   I. Title.

RA591.S83   1990   628.1'61   87-24315

ISBN 0-943163-16-1
ISBN 0-943163-15-3 (pbk.)

With love to Sharon (the missus) for:
her faith in me and the project,
which often surpassed my own;
her continual support and encouragement,
unparalleled in the known world;
and her A to Z system
that got me from A to B.
She makes each day bright and hopeful, and helps
turn dreams into reality.
Without her, this book would not exist.

Acknowledgements

It is impossible to undertake a project of this magnitude without the assistance, input, and encouragement of family, friends, and noted experts. Special thanks to the following eminent professionals for their time, wisdom, and insight in the technical review of portions of the manuscript: Dr. Leon F. Burmeister, Professor and Vice Chairman of Preventive Medicine and Environmental Health at the University of Iowa; Kristie Thorp, Conservation and Environment Project Leader, Center for Rural Affairs, Walthill, Nebraska; Dr. Dennis Taylor, Professor of Biology, Hiram College; C. Richard Cothern, Ph.D., Science Advisory Board, U.S. EPA; and John A. Yiamouyiannis, Ph.D., President of the Safe Water Foundation. Be advised that the opinions expressed in this book are those of the author and do not necessarily reflect those of the reviewers.

Innumerable thanks for their time and effort to: Bobbie Martin; Victor and Anna Maffe; Daniel Stewart, Esq.; Dr. James Stewart; Ellen Stewart; and Michael Lezark for proofreading and helpful suggestions. Very special thanks to my parents, Delores and Eugene Stewart, Esq. for literally all forms of assistance, encouragement, and rescue. Mammoth quantities of gratitude and major league thanks to John and Becky Cartus for coming in at the last minute and saving the entire project from the undertow of swirling potty water. Their world class editing, organizational skills, and command of the language changed the manuscript from a rambling mess into the book that lies before you. Some measure of appreciation must also go to Spanky, a loyal and constant companion during incalculable hours of research and writing. He never let me down despite the neglect that accompanied the years of work on this book.

Table of Contents

Warning — Disclaimer

The purpose of this book is to act as a resource of information for those concerned about the quality of the water they drink, but only within the confines of the material covered. The sale of this book is accompanied by the understanding that both author and publisher are not giving legal, medical, or any type of consulting advice. As every situation is different, one solution cannot be universally applied to all water quality or contaminant exposure problems. This book is not a substitute for direct professional consultation. Legal, medical, or technical advice regarding the quality of the water routinely used for drinking and other purposes should be obtained by direct interaction with competent professionals.

Considerable effort was made to make this book as accurate and comprehensive as possible based on information generally available up to the publication date. None-the-less, both typographical and content errors may exist. Consequently, this manual should be used as a general guide and not the sole source of information on the assessment of home water quality.

The intent of this book is to inform and entertain. The publisher and the author shall not accept liability nor responsibility to any person, persons, or entity with regard to any loss or damage what-so-ever caused or allegedly caused by information or omission of information from this book.

Warning — Disclaimer

INTRODUCTION

"Writing a book is an adventure. To begin with, it is a toy and amusement. Then it becomes a mistress, then it becomes a tyrant. The last phase is that just as you are about to be reconciled to your servitude, you kill the monster, and fling him to the public."

— Winston Churchill

In the past, most people gave little thought to the source of their water, its vulnerability to pollution, or its overall quality. Water was taken for granted. Many people saw water as something that mysteriously gushed from their faucet and believed that some all-knowing power assured its purity. In recent years, however, several factors have contributed to a greater concern about drinking water quality. They include better scientific detection methods, which have found contaminants at lower levels than previously thought possible, a greater health and environmental awareness by the public, and an increase in reports of drinking water contamination.

Generally, people are becoming more aware and are taking steps to assure good health. They are exercising, eating the right foods, minimizing chemical exposure in the workplace, testing their homes for radon, and having their water analyzed. Today, people want to know. They want to be informed about things that affect them, particularly when it concerns their health. Drinking water is no exception, although it may be overlooked by those who are unsure of what to do. In the same way, people generally still follow the cliché, "my home is my castle." Just as one secures a home against break-in, and spends money on home improvements, it only makes sense to invest in continued good health by assuring a safe home water supply.

During my years as a laboratory chemist, many people, like yourself, concerned about the quality of their water, came into the lab wanting to know how to go about testing their water, what to test for, what the test results meant, etc. To answer these questions and offer a guide in determining home water quality, I wrote **Drinking Water Hazards.** This book will guide you in determining if contaminants exist in your water and provide you with the information you need to make decisions. I will discuss common pollutants and their potential sources, how to choose a laboratory to test your water, standard tests you can request, interpreting the test results, and some home treatment and legal options should you find contamination. We also look at how we got to the point where contamination of water is a widespread problem, take a peek at where we are going, and present some things you can do to make a difference in the future. The book is designed as a handbook and a ready reference with chapter subheadings to help you to easily locate the information you want.

Determining the status of water quality can be a difficult task, even for professionals. However, it is by no means impossible and should not be completely out of your hands. It is your water, and if you are concerned enough to explore its quality, you should be actively involved in the process of testing it. No one is looking out for you but you. Armed with the necessary information, you can make sound decisions which directly affect your family's health.

Bringing any book to press is a complicated process and it is certainly possible that errors exist. If you discover what you feel to be an error, or have any suggestions for improvements to this book, please send them to me c/o Envirographics, P.O. Box 334, Hiram, Ohio 44234. Regretably, time constraints do not allow me to answer all correspondence, but know that I appreciate your assistance. Feedback and direct communication are essential to the production of any informative publication.

— John Cary Stewart

1. ♦ WHAT IS WATER?

"For he hath founded it upon the seas, and established it upon the floods."

— Psalms 24:2

Roughly 75% percent of the earth's surface is covered with water. According to the theory of evolution, life began in water. When these small microorganisms evolved into larger complex organisms and moved onto the land, water went along as a significant percentage of their bodies. Water comprises about 70% of the human body. It is the basis for life in any ecosystem because it directly supports the first links in the food chain. Microorganisms living in surface waters are food for larger invertebrates, which are food for fish, which are food for mammals like humans. On earth then, water *is* life.

Proximity to surface water sources determined where early man chose to live. Valleys were often the bases for early civilizations. When he began to farm, man needed water to irrigate his crops. During the industrial revolution man built his factories beside waterways because water was needed for manufacturing processes.

Water is more than what many see as simply H_2O. It is a unique molecule with unique properties. Water has been called the universal solvent because its configuration breaks apart many compounds and also prevents them from recombining. In this way, water dissolves natural chemicals and nutrients necessary for life's processes so they can be easily absorbed by, and carried throughout, the human body. It is believed that many beneficial and necessary minerals and nutrients are better absorbed by the body from water than from foods. In addition, water is instrumental in flushing the body of wastes.

Water's excellent solvent capability also makes it easily polluted if pushed beyond its natural ability to cleanse. Today, many waterways are overloaded beyond their ability to biodegrade wastes. In addition, many of the chemicals created by man are not water soluble. These chemicals are very persistent, do not readily break down, and therefore disrupt natural chemical and biological cycles within water and surrounding ecosystems.

Water also has a high **specific heat** and **heat of vaporization** which means that it stores heat very well. Large amounts of heat are necessary for water temperature changes. As a result, warming in the spring and cooling in the autumn occurs slowly. In addition to maintaining a steady water temperature and preventing rapid and wide fluctuations which could shock aquatic organisms, this also stabilizes the temperature of entire geographic areas. Uncharacteristic of many other liquids, water expands when it freezes and has its maximum density above its freezing point. This allows ice to float and prevents large bodies of water from freezing completely so aquatic life survives below the ice during the winter.

The **Hydrologic Cycle** (Figure 1-1) is the natural pathway water follows as it changes between the liquid, solid, and gaseous states. The continuing process of evaporation, condensation, and precipitation moves water between the five portions of the hydrologic cycle: oceans; water vapor in the atmosphere (clouds); ice and snow; groundwater; and surface water lakes and streams. The hydrologic cycle is a complex movement of water in the atmosphere, on the land surface, and underground. It determines precipitation activity and replenishes both ocean and freshwater sources throughout the world.

Water vapor can travel large distances before it recondenses as precipitation — the only source of water for the earth's freshwater supply. Precipitation falling to the earth's surface either filters through the soil into groundwater or drains along the land surface as runoff to surface streams or lakes. The hydrologic cycle is a natural water treatment process. It is effective as long as the water in it is not contaminated beyond the capacity of the cycle to cleanse. An imbalance in this cycle due to man's activities can have devastating effects.

Of all the water found in or on the earth, 97% is found in the oceans.

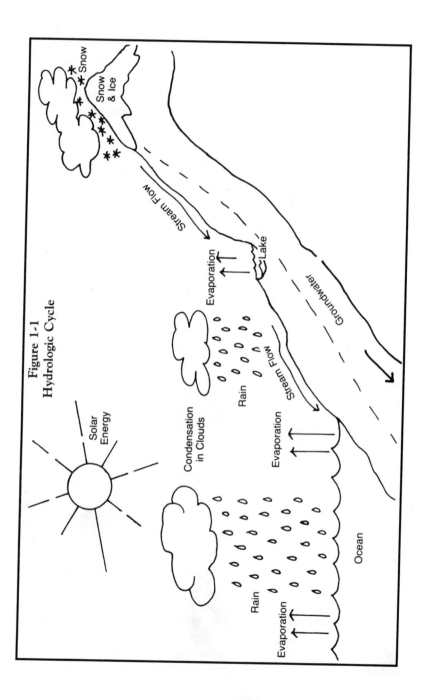

Figure 1-1
Hydrologic Cycle

The remainder is fresh water. However, less than one-half of a percent (0.5%) is available in rivers, lakes, or accessible groundwater sources. When one considers the amount of water too heavily polluted for use, or those sources too expensive or troublesome to utilize, we are left with only 0.003% of the world's water supply. To put this in perspective, if all the water in the world was represented by a 55 gallon drum, the readily available freshwater would fit in a half-quart bottle, and the usable unpolluted freshwater would be represented by half a teaspoon. However, this amount, if used wisely and not degraded irreversibly by humans, is still considered more than enough for the world's current inhabitants, particularly since the hydrologic cycle continually purifies and disperses the water.

Groundwater is utilized by 50% of Americans, yet it is poorly understood by most of them. Groundwater is recharged mainly by precipitation which drains through the overlying soil, but can also be recharged by surface waterways that contact underground waterways. Gravity carries the water through the upper soil layers into porous rock or sand and gravel layers with water flowing through them. These waterways are called *aquifers.* The relatively slow groundwater recharge process, compared to the rapid runoff to surface waters, must be considered in the use or overuse of the groundwater resource to the point where it is non-renewable.

Groundwater and surface water, the main sources upon which man depends, have very different characteristics. Surface water supports a wide array of plant and animal life, including microorganisms like bacteria and algae, which depend on the prevalence of dissolved oxygen, nutrients, and biodegradable matter. Surface waters are a web of life that can be seriously affected by pollution from sewage or synthetic chemicals. Groundwater, on the other hand, is usually low in oxygen. Microorganisms are also low in groundwater because the soil filters them out of the water as it drains through to the aquifer. Groundwater is, however, higher in minerals dissolved from rock formations through which it travels. Sometimes the mineral content of groundwater may be undesirably high. Iron, chloride, or salts can be elevated in certain areas.

Specific sources of groundwater pollution are discussed in Chapter 14. These are often easier to pinpoint because surface water contaminants mix together making it difficult to identify a specific discharge point as being the pollution source. A river or watershed system can have such a large industrial and sewage discharge base that it would be nearly impossible to trace a chemical or chemicals to a specific source. Groundwater moves much more slowly than surface water and, when polluted, contaminant concentrations are often much higher than in surface water. Consequently, it is easier to detect contaminants.

Water Contamination

Potential water pollutants are more numerous and varied than you might think. For ease in discussing them, they can be classified as:

Biological Contaminants — These are parasites, bacteria, viruses, or other undesirable living microorganisms that enter the water mainly via human sewage and proliferate to levels that can cause disease when ingested through drinking water.

Inorganic Chemicals — Chemicals that occur naturally but have been mined, processed, refined, and concentrated by man, to the point where they have become contaminants. Examples include: cyanide, fluoride, and heavy metals like lead.

Radioactive Elements — They exist in nature at low levels, but nuclear power plants and military installations are responsible for adding additional radioactivity, often at higher levels than are found naturally in waterways.

Fertilizers — Nitrogen and phosphorous from agricultural fertilizers and sewage are plant nutrients that cause an overabundance of bacteria and algae in lakes when high amounts of this "food" are available. Nitrate (nitrogen) in particular is a potential problem in drinking water.

Synthetic Organic Chemicals (SOCs) — Chemists define an organic chemical as one that contains carbon. This can be a bit confusing because many people associate the word "organic" with nature or natural products. This is *partly* true. Organic chemicals are everywhere in nature, trees, plant life, soils, and in humans. Interaction of these chemicals are the basic processes of life and the life cycle. Synthetic organic chemicals are those which are created by man in the laboratory. The bottom line is that both natural and manmade chemicals contain carbon. However, many of those synthesized by man are dangerous pollutants and have no place or niche in waterways or any other part of the environment. They are foreign to nature and are a serious potential detriment to it, humans included. Seriously damaging health effects have been linked to many SOCs. Examples of familiar SOCs include: industrial solvents, like TCE; pesticides, like DDT; PCBs; and dioxins. (SOCs are discussed in greater detail in Chapter 12.)

Several constituents or physical properties of water are also helpful in understanding some aspects of water pollution. They are mentioned throughout the book and in some cases may have a bearing on the availability or level of certain contaminants. A brief discussion of each now will be helpful and act as a point of reference:

Acidity and pH — The pH of water is easily determined by dipping pH paper into the water and observing a color change, or by use of a pH meter in the laboratory. The pH of water will indicate how acidic or alkaline the water is. A pH scale of 0 to 14 is used to indicate the varying degrees of acidity, with 7 being neutral (Figure 1-2). Anything below 7 is acidic, anything above is alkaline. So, the lower the pH, the higher the acidity. Water that is acidic is no problem in itself. The problem is that the more acidic the water, the more corrosive it is. Corrosive water strips household plumbing of metals like lead which can cause serious health effects (see Chapter 7).

Turbidity — Turbid water contains suspended particles to the extent that it becomes cloudy and interferes with the ability of light to pass through the water. The suspended particles associated with turbidity are due to natural silt, clay, soil, decaying vegetation, microorganisms, and industrial waste discharge. One of the main problems with turbidity is the fact that toxic chemicals discharged into waterways often attach to these suspended particles (see Chapter 8).

Hardness — Hard water contains a high amount of calcium, magnesium, or iron in addition to other minerals. Problems with hard water have traditionally been economic or aesthetic, such as the reduction in the cleansing power of soap and the build up of scale in hot water heaters, boilers, and hot water pipes. However, while many seek to remove these minerals by softening their drinking water, an increasing pool of evidence points to the lack of mineral availability as contributing to certain health effects (see Chapter 6).

An explanation of how levels of contaminants are given is necessary to help you understand some of the concepts discussed in subsequent chapters. Levels of contaminants in water are expressed as the weight of the contaminant in a volume of water, usually liters. And yes, unfortunately for those non-scientific residents of the United States, these levels are **always** given in metric units. However, this need not send you screaming

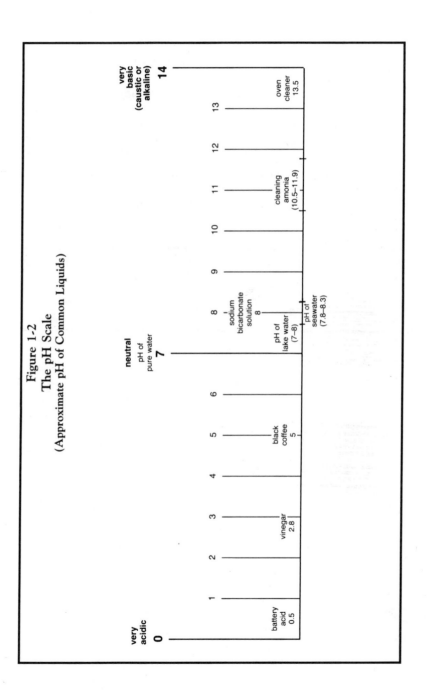

Figure 1-2
The pH Scale
(Approximate pH of Common Liquids)

into the night. The problem is that most people try to convert from metric to standard U.S. units in order to understand, or get clear in their minds, the amount being discussed. Keeping the amount or level in metric units and using a common comparison is much easier to comprehend. For example, 1 milligram per liter (1 mg/L) means there is 1 milligram of a contaminant in 1 liter of water. This is the same as one part contaminant to 1 million parts of water, and is therefore called one part per million (1 ppm). To get an idea of what 1 ppm is, one drop of food dye in a bathtub of water would be 1 ppm. Some contaminants like synthetic organic chemicals (SOCs) are measured in the parts per billion (ppb) range due to their toxicity at lower levels. Table 1-1 puts all these in perspective and gives conversions.

Table 1-1

1 milligram per liter (1 mg/L) = 1 ppm
1 microgram per liter (u g/L) = 1 ppb
1000 ppb = 1 ppm

---

2. LIVING IN A CHEMICAL WORLD

"While we spend our life asking questions about the nature of cancer and how to prevent it, society merrily produces oncogenic substances and permeates the environment with them. Society does not seem prepared to accept the sacrifices required for effective prevention of cancer."

— Reenato Dulbecco,
1975 on acceptance of the Nobel Prize
in Medicine for research on cancer

One hundred years ago man was merely on the doorstep of the current technological age. The recent end to the Indian Wars in the west marked the end of the frontier. Life had not changed remarkably for mankind up to this time. The industrial revolution had brought him to the threshold of the 20th century, but the changes that had occurred since recorded time were dwarfed by those yet to be realized in the coming century.

In the decades since the Second World War, the United States has experienced previously unseen technological and economic growth. Unquestionably, man has benefited from the advances of this period, but this path of progress has left in its wake a myriad of measurable, as well as yet unrecognized, detrimental effects. Most people alive today were born into, and grew up within, this period of unprecedented growth and development. We have come to accept and even welcome new unchecked innovations and advances, including an ever increasing number of synthetic chemicals. The postwar era marked a change from products derived from natural materials, which can be easily degraded by natural systems, to synthesized products which do not easily biodegrade (Table 2-1). Annual production of synthetic organic chemicals (SOCs) increased from approximately one billion pounds in 1940 to over 387 billion pounds today.[1]

Table 2-1
Some of the Synthetic Products That Replaced
Natural Products After 1950

Natural Products Used in 1950	Synthetic Products Used Today
Animal and Plant Fibers (Cotton, Wool, Silk)	Synthetic Fibers
Wood	Plastics, Steel and Aluminum
Soap	Detergent
Pure, Natural Foods	Foods with Additives and Preservatives
Organic Fertilizer (Manure)	Commercial Chemical Fertilizers
Natural Predators	Synthetic Pesticides
Rubber	Synthetic Rubber

Adapted from: LIVING IN THE ENVIRONMENT: AN INTRODUCTION TO ENVIRON-MENTAL SCIENCE, 6/E, by G. Tyler Miller, Jr. ©1990, 1988 by Wadsworth, Inc. Used by permission of the publisher.

Effects of Living in a Chemical World

Although arguably increasing our quality of life, modern technology has also brought with it certain inherent hazards. Human endeavors have reached the level at which large natural ecosystems that have endured and evolved since the beginning of time are threatened. In terms of health effects, humans have not escaped these negative consequences. The potential health effects of living in the modern age are innumerable and ominous.

Up through the early part of this century, infectious diseases like influenza, pneumonia, and tuberculosis were topping the list of deadly human afflictions. By mid-century the figures pointed to a drastic decrease in infectious diseases, but a clear increase in diseases like cancer, strokes, and heart disease. Cancer was the eighth leading cause of death in 1900. Today it is second.[2] In 1987 an estimated 965,000 people developed cancer, and 30% of the Americans alive today will eventually develop some form of cancer. Cancer kills more children between the ages of three and 14 than any other disease.[3]

Overall cancer incidence rates have risen sharply in the post-war era.[4] The Toxic Substances Strategy Committee in its 1980 report to the President concluded that both cancer incidence and cancer mortality are increasing even after considering both age and the effects of cigarette smoking.[5] Between 1950 and 1985, overall cancer incidence rates increased 37%.[6] For some specific organ sites, the incidence increase is much higher: 100% for non-Hodgkins lymphoma, multiple myeloma, and malignant melanoma; 92% for testicular cancer; 67% for prostate cancer; 63% for colon and rectal cancers; and 142% for kidney cancers in males.[7] In addition, testicular cancer rates in white males has doubled, and in black males tripled, since 1950. A hundred years ago testicular cancer was almost nonexistent in men under 50. Today for men ages 15 to 34 it is one of the most common cancers.[8] Cancer mortality accounted for one death in six in 1968. In 1985, it accounted for about one death in five.[9] According to the National Cancer Institute (NCI), environmental factors, including lifestyle considerations like smoking, diet, and synthetic chemical pollutant exposure, are responsible for approximately 90% of cancer incidence.[10] Therefore, many cancers could be avoided by the betterment of our environment.

Some scientists believe that our increased exposure to carcinogenic chemicals, as a result of the sharp escalation in their production after World War II, is responsible for a measurable portion of these increases.[11] Variations in cancer rates have been observed based on geographic area. Higher rates of mortality from lung cancer and cancers of the respiratory tract and skin were found in male residents of counties with industries involved in petroleum production.[12] A study of lung cancer mortality in relation to distance between residence and industrial plants concluded that of the 13 industries studied, the petroleum and chemical industries exhibited the greatest risk.[13] The results were shown to be independent of the occupation of the people studied. Researchers have reported a positive relationship between estimated levels of exposure of residents to air emissions from chemical plants and incidence rates for a number of cancers.[14] A 1989 report by the Natural Resources Defense Council identified 1,600 major source sites nationwide emitting eleven carcinogenic, synthetic organic chemicals into the air.[15]

Another study of cancer mortality by county found bladder cancer in males pointed an accusing finger at chemical, auto, and machinery manufacturing. Salem County in New Jersey led the nation in bladder cancer deaths for white men. The researchers concluded this excess risk was due to occupational exposure as 25% of the workers in the county were employed by the chemical industry.[16] According to the National Center for Health Statistics, bladder cancer death rates are highest in the heavily industrialized northeastern United States.[17] A geographic study of cancer mortality in counties with chemical industries found increased bladder, lung, liver and other cancers in white males in 139 counties where chemical companies are most heavily concentrated.[18]

An 85-mile stretch of the Mississippi River between Baton Rouge and New Orleans is known as "cancer alley" because of the higher rates of certain cancers. Louisianna itself has the highest lung cancer death rate for white males in the country. For black males the lung canccer death rate is the highest in the world. Some residents also report a significant number of miscarriages. The river segment is lined with seven oil refineries and 136 petrochemical plants.[19]

Chemicals and Reproductive Abnormalities

Reproduction, whether in humans or animal species, is an extremely complex and highly vulnerable physiological process. All process actions must proceed uninterrupted and in sequence for the miracle of life to take place. Many points exist along the way for interruption or devastating alterations. Very little is known about the causes of reproductive abnormalities such as infertility and birth defects. Birth defects research remains generally in the data gathering stage. According to a report by the Council on Environmental Quality, the current knowledge of toxic chemicals and reproduction is similar to what was known of chemical carcinogens in the late 60s when the impact of chemicals as carcinogens was only beginning to be realized.[20]

Most information accumulated on reproduction is solely for routine medical data bases, as opposed to environmental effects monitoring. Many reproductive effects are difficult if not impossible to assess. Impotence, infertility, and spontaneous abortions are not ususaly recorded or documented. Birth certificates only record the most severe or apparent abnormalities.[21] Despite the fact that chemical effects on reproduction cannot be discerned from this information, patterns of reproductive performance can be seen.

- Three out of every 100 human babies will come into this world with major defects.[22]
- According to one researcher, an estimated 23–35% of birth defects have a genetic foundation, while 7–11% are due to environmental

factors.[23] However, in the remaining 65–70% of the birth defect cases, no cause could be specifically identified.

- An estimated eleven million married couples (15% of the total) in the United States are infertile.[24]

Female reproductive system development is believed to be particularly vulnerable to chemical substances. Fertility reduction in the rising generation is believed to be indicative of prenatal exposure to chemical toxins.[25] Researchers found over 100 halogenated hydrocarbons, a class of manmade chemicals, and plastic components in the umbilical cords of newborn infants.[26] Some of the chemicals were found at levels above those found in the mother's blood, indicating the movement of the chemicals to the fetus is not reversible. These include common chemicals like benzene, carbon tetrachloride, and chloroform.

Only in a very few instances is it possible to determine the reproductive effects of exposure to a particular chemical. Occupational exposure of females to certain chemicals has been implicated in reproductive and genetic health effects. A study of chemical workers in Finland found an increased rate of spontaneous abortion, specifically those women in plastics and styrene manufacturing.[27] Another Finnish study determined that risk of spontaneous abortion was significantly increased if mothers were employed in industrial, agricultural, and construction jobs rather than in office, bookkeeping, and homemaking occupations.[28] Other researchers reported significant increased incidence of serious malformations[29] and chromosome abnormalities[30] in children of female laboratory personnel.

Chemical exposure may also be a factor in male infertility. Ralph C. Dougherty, professor of chemistry at Florida State University discovered a possible link between chlorinated hydrocarbons and a decrease in male fertility.[31] According to his research, male fertility, as determined by sperm densities, has decreased 80% since 1929. His study of college age students found about one in four had a low sperm count. All samples had SOCs in them, including PCP, PCBs, and breakdown products of DDT. Halogenated organic chemicals, synthetic chemicals containing chlorine or bromine, are of the most concern. The pesticide DBCP (dibromochloropropane), for example, sterilized, at least temporarily, nearly all men working with it at a California factory in the 1970s.[32]

The legacy of chemical mutagens may be even more ominous and horrifying than those of carcinogens. The latency period between exposure to a carcinogen and the onset of the disease is measured in decades. However, the mutagenic effects of current chemical exposure may not be known for several generations. In the cost/benefit breakdown of the usage of a specific chemical, often each is allocated to different segments of society. The health and environmental effects vs. the short-term increased productivity of pesticides is a good example. But in the case of chemical mutagens, the benefits are secured by those alive today, while the cost is shouldered by those that may not be born for another century.[33]

Pollution and Health Effects

In certain areas of eastern Europe the health effects of environmental contaminants are sharply pronounced. With its heavy industrialization, large per capita use of resources, and meager committment to pollution control, Poland is in the midst of an economic and ecological nightmare. It stands as a graphic example of the consequences of uncontrolled industrialization and pollution. Nearly half of the nation's rivers are not fit even for factory use.[34] Only one-fourth of the water pumped from public wells is classified as "good."[35] Less than half of Poland's 800 cites have sewage treatment plants. Warsaw is scheduled to get one this year.[36] One quarter of the soil in Poland is not suitable for food production.[37]

Over 30% of the residents of Poland will likely develop a disease as a result of environmental pollution — respiratory disease, cancer, etc.[38] Life expectancy, while increasing in certain parts of the globe due to advances in health care, is decreasing for certain age groups in Poland.[39] In the heavily industrialized Upper Silesia region, the cancer rate is 30% higher and there are 47% ·more cases of respiratory illnesses than the national average. In Upper Silesia there has also been a dramatic increase in the number of retarded children.[40] Just over the border in Czecheslovakia's northern Bohemia, life expectancy is ten years less and skin disease, stomach cancer, and mental illness are double the national average.[41]

We are becoming increasingly aware that seemingly innocuous human activities can cause severe environmental damage and ultimately human health effects. But, we are only beginning to realize the magnitude of these effects:

• Strong evidence of a link between aluminum in the environment and Alzheimer's Disease has been proposed. British scientists determined that people who drink water with elevated levels of aluminum were at least 50% more likely to develop Alzheimer's Disease than those exposed to very low levels of aluminum in water.[42] Aluminum has been found in the brains of Alzheimer's patients and its ability to get there may be related to its bioavailability.[43] While aluminum is a common element to the environment, historically it has not been easily absorbed by plants and animals. It is speculated that acidic deposition of air pollutants, like acid rain, may be one of the factors contributing to the uptake of aluminum by biological systems. A noted effect of acidic deposition is the leaching of aluminum and other metals from soil that does not have the buffering capacity to withstand the acidic rain. In the recent time that acid deposition has become recognized as a problem, increased amounts of aluminum have been found in natural ecosystems with damaging effects to plant and animal life. Acid rain may also dissolve metals and other contaminants from water supply distribution systems.[44] The total effects of acid rain are not fully known, including the effects on drinking water quality.

- Parkinson's disease incidence, virtually unknown prior to the industrial revolution, but rising steadily through the early part of this century, was found in a Quebec study to strongly correlate with the level of pesticide use.[45]
- In the wake of studies pointing to an increased incidence of leukemia in the area near a nuclear power plant in Plymouth, Massachusetts, and similar observations in Britain, the National Cancer Institute is studying cancer deaths around over 100 nuclear power reactors in the U.S.[46]
- Elevated rates of leukemia deaths have been seen in ten states in the central United States.[47] According to the National Center for Health Statistics, this is believed to be the result of occupational exposure of farmers to agricultural chemicals like pesticides.
- A study of pesticide use in and around the home concluded that regular use of pesticides in the house translates to almost four times the likelihood of leukemia incidence in children living there. With the regular use of garden sprays, the risk of children in the household developing leukemia is over six times what it would be if the sprays were not used.[48]
- Several studies have determined a link between parents' occupation and cancer in children. A parent's exposure to chemical hydrocarbons during employment hours is believed responsible for an increase in risk of childhood leukemia. Researchers have found a significantly increased risk of leukemia in children whose fathers are exposed to chlorinated solvents, spray paint, dyes, pigments, and cutting oils.[49]

Our lifestyle and use of certain chemicals also has ominous implications on the global scale as well. A rise in the levels of atmospheric carbon dioxide (CO_2), nitrogen oxides, and methane, which do not allow infared radiation and heat to escape from the earth's surface, are responsible for what scientists call the ***greenhouse effect.*** The increase in carbon dioxide and other greenhouse gases is the result of fossil fuel combustion from sources like automobile exhaust and industrial emissions. Fossil fuel combustion has resulted in a rise in the CO_2 level in the atmosphere from 280 parts per million (ppm) prior to the industrial revolution, to 348 ppm in 1987.[50] Satellite measurement of global ocean surface temperature shows a significant increase, about 0.1°C yearly, between 1982 and 1988.[51] Average global temperatures increased approximately 0.6°C in the last 100 years.[52] If the current rate of CO_2 emissions continues, scientists predict a 2.5° to 5.5°C rise in global temperature in the next century.[53] This would result in dramatic climate changes, disruptions in agricultural production, and large scale refugee problems. While it will take another ten years to be certain if a clear trend exists, the 1980s had six out of the ten warmest years ever recorded, making it the hottest decade in global temperature measurement history.[54] Some scientists foresee a continuation of this trend in the years to come.

It was originally thought that abundant forest plant life would absorb this excess carbon dioxide through photosynthesis — the exchange of oxygen for carbon dioxide. However, while carbon dioxide levels continue to increase, global forest area is being reduced by widespread clearing and burning. Burning not only reduces forest area, but produces yet more carbon dioxide.

Chlorofluorocarbons (CFCs), a class of chemical compounds once heralded as wonder chemicals for their wide range of uses, are now blamed for ozone depletion in the earth's stratosphere. Ozone absorbs dangerous ultraviolet (UV) radiation emitted by the sun. Losses in stratospheric ozone causes more UV radiation to reach the surface of the earth and affect humans in terms of significant increases in eye damage, skin cancers, and injury to the immune system.[55] Between 1979 and 1987, the levels of ozone in the stratosphere have fallen, and a large hole twice the size of the United States has been discovered in the ozone layer over Antarctica.[56]

Small microorganisms in the oceans can be irrevocably harmed by UV radiation, resulting in widespread disruption of natural food chains. Crustaceans eat these microorganisms, which are in turn eaten by small fish, which are eaten by larger fish, which are eaten by whales and humans.

The CFC concentration in the atmosphere has increased despite bans on aerosol sprays which use them as a propellant. The current level of chlorine from CFCs in the atmosphere is reportedly five times what it was in 1950.[57] Approximately 650 million pounds of CFCs were produced in 1985 in the United States, slightly less than a third of the total world production. Plans are in place to reduce and eliminate CFC production in the U.S., however, a quick global agreement to phase out CFCs is necessary because they stay in the atmosphere for more than 100 years. In addition to harmful UV radiation, CFCs also absorb heat radiated from the earth's surface, not allowing it to escape into the atmosphere. Consequently, CFCs are also gases that contribute to the greenhouse effect.

The Staggering Hazardous Waste Problem

In 1950 when SOC production was in its infancy, the annual amount of hazardous waste generated was about 5 pounds for every person in the country. Today, at an annual production rate of about 292 million tons, it is a staggering 2,500 pounds per capita annually.[58] The large scale production of synthetic chemicals for a wide variety of industrial uses and consumer products, coupled with traditionally haphazard disposal practices, has resulted in an enormous hazardous waste problem. Decades of mismanagement by those generating, storing, treating, using, and disposing of hazardous chemical wastes has brought us to the point where possibly hundreds of thousands of sites nationwide are leaking hazardous chemical wastes into groundwater and surface waters used for drinking.

In 1980, the Comprehensive Environmental Response, Compensation, and Liability Act (CERCLA), better known as Superfund, was enacted to give the EPA authority and money to take corrective measures and clean up hazardous waste sites. The 1986 Superfund Amendments Reauthorization Act (SARA) outlined preferred cleanup methods including permanent, on-site treatment. However, several reports have concluded that Superfund is not being correctly implemented. According to a Congressional Office of Technology Assessment (OTA) report, the EPA is using essentially cheap, stopgap methods, such as "capping," at hazardous waste sites rather than permanent treatment or destruction methods.[59] Often wastes are simply moved to another site which may eventually leak as well. A report by the Senate subcommitee on Superfund, Ocean and Water Protection concluded that ". . . EPA's implementation of Superfund has, in general, failed to meet public expectations and mandates of the law."[60] In addition, the report concluded that the EPA's enforcement actions have resulted in cleanup strategies that in many cases are *not* "achieving the legally mandated goal of permanent treatment."[61] A survey by the House Subcommittee on Oversight and Investigations in 1985 determined that 54% of the land disposal facilities (including landfills and surface impoundments) which receive Superfund waste lack adequate monitoring to determine leakage of waste into groundwater supplies.[62]

A 1988 report by several environmental public interest groups and the Hazardous Waste Treatment Council, **Right Train, Wrong Track: Failed Leadership in the Superfund Cleanup Program,** accuses the EPA of capping and containment remedies at hazardous waste sites.[63] These methods will require additional and complete cleanup sometime in the future when the cost will be much higher. Additionally, it does not elimi- nate the hazards that permanent treatment would. According to the report, in fiscal year 1987, cleanup methods at Superfund sites did not utilize treatment of contamination sources in 68% of the cases studied. The researchers concluded that the EPA:

> "set cleanup goals unscientifically, exempted Superfund from the very environmental regulations it imposes on other waste manage- ment facilities (i.e., hazardous waste disposal restrictions and liner requirements for landfills), and ignored the impact of Superfund sites on natural resources in the vast majority of its cleanup deci- sions."[64]

The EPA, due to its limited resources, concentrates its authority on the hazardous waste sites on the National Priority List (NPL). This list of the most severely contaminated sites is continually growing. As of March 1989, 1,163 sites were listed on the NPL.[65] However, the EPA Comprehensive Environmental Response, Compensation, and Liability Information System lists 30,844 sites that may require addition to the NPL.[66] Only 26 sites have been removed from the NPL since the ratification of Superfund.[67]

The average Superfund site takes six to ten years to clean up and costs an estimated $18 to $30 million.[68] But, clean-up costs continue to rise. In 1986, Superfund was reauthorized for $8.9 billion. However, esti- mates of the total cost to clean up the known hazardous waste sites are as high as $40 billion.[69]

As of March 1989, there were 1,451 hazardous waste disposal sites covered under the Resource Conservation and Recovery Act (RCRA). According to the GAO, over 70 percent may be leaking and more than half may require measures to reduce groundwater contamination.[70] The EPA estimates that the extent of corrective action needed to repair and contain leakage at RCRA sites may rival that of the Superfund sites.[71]

The EPA's 1983 Surface Impoundment Assessment National Report lists 181,000 pits, ponds and lagoons — collectively called surface im- poundments — that accept either hazardous or nonhazardous waste.[72] Industries using this mode of "waste treatment" include the chemical industry, food processors, mining operations, paper industries, and metal processing. In the past, surface impoundments, like landfills, and other waste sites, were situated and used without consideration of the impact on groundwater quality. Fifty percent are over top-soil layers that are either narrow or permeable to pollutant movement to groundwater. Seventy percent are above aquifers that can easily spread a contaminant.[73] More than 98% are located within a mile of drinking water or potential drinking water sources.[74] An embarrassing 1,564 (less than 2%) of these sites have monitoring wells to detect leakage to groundwater.[75]

According to a GAO report, "a complete inventory of hazardous waste sites does not exist." As the EPA points out, if discovery programs were initiated many additional sites would be added to the inventory. The GAO observed that the EPA generally has been concentrating on cleanup of known sites rather than searching for new ones, and does not have a plan for identifying new sites in the near future.[76] The GAO estimates that anywhere from 130,300 to 425,000 sites may need to be evaluated and included on the Superfund NPL.[77]

The lack of a complete inventory means the extent of the hazardous waste problem in this country is not known. People living near unknown or undiscovered waste sites are unaware of the dangers and may be unwittingly exposed to a wide variety of chemicals in the air and water for a number of years.

Another problem is the uncertainty of the sources of hazardous wastes. The GAO reports:

"We cannot estimate the amount, location, and source of hazardous waste being produced, either nationally or at the state level with confidence."[78]

An OTA study echos these concerns:

"Data inadequacies conceal the scope and complexity of the nation's hazardous waste problems, and impede effective control."[79]

As a result, we do not know whether treatment, storage, and disposal capabilities will be adequate to handle hazardous waste volumes; or the extent of illegal disposal of hazardous waste. The GAO concludes that Congress lacks the data base to effectively plan for the management of the nation's hazardous waste. This transfers to uncertainty about the level of toxic chemical exposure to humans through mediums like drinking water.

Sadly, the seemingly insurmountable problem of dealing with past mismanagement of our hazardous wastes is only half the battle. We are still not dealing effectively with the wastes currently being produced. Because there is an absence of adequate means of treating hazardous waste, the EPA has been slow to implement legislation passed in 1984 intended to restrict land disposal of hazardous waste. Also, minimizing the production of hazardous waste at the industrial source is not being encouraged to any great extent through legislation or government policy. Consequently, we are creating new waste problems while we wrestle with old ones, thereby putting ourselves into a perpetual cycle of cleaning our toxic chemical messes.

The Legacy of the Chemical Age

We, as humans are experiencing what all the natural world is experiencing — the effects of the chemical age. We have been introducing damaging chemicals to the world's ecosystems, particularly in the last century. In that time we have seen widespread environmental degradation, both nationally and on a global scale, as well as a rise in cancer rates and other diseases. But what about the health effects we are unaware of or have yet to discover? Samuel S. Epstein, M.D., professor of Occupational and Environmental Medicine at the University of Illinois, and an internationally recognized authority on the toxic effects of chemical pollutants, summed up the problems of chemicals in the environment:

> "We are living in an era of organic chemicals, not just familiar ones, but exotic ones which have never previously existed on earth, and to which no living thing has previously had to adapt."[80]

We are currently in the middle of a large experiment on human exposure to hazardous chemicals of our own making. Like it or not, all of us are part of this experiment. While much is known, many pieces of the puzzle may not be available for decades, if ever.

3. WATER POLLUTION & DRINKING WATER CONTAMINATION

"The products of population growth and industrial development are reflected in the deteriorating quality of our drinking water in many areas across the country."

— Charles C. Johnson Jr.
in a speech before the
American Medical Association, 1983

At any given time in the history of mankind, the quality of water sources has been directly dependent on the human population level and the methods of waste disposal used by that population. Despite man's lengthy misuse of the water resource, a conclusive link between water and disease was not proven until the widespread cholera epidemics of the 1880s. Efforts to combat cholera, typhoid fever, and other waterborne diseases led to the establishment of comprehensive biological water treatment practices which remain in use today.

It is not surprising that the advent of synthetic chemicals also brought about additional degradation. Water; as the eventual drain for much of man's industrial waste, did not escape this chemical onslaught. As precipitation washed land disposed waste into waterways, and industrial sites and sewage treatment plants discharged effluent directly into rivers and streams, the contamination of drinking water logically followed.

The reasons for the current state of water and drinking water quality are easily defined:

- *population growth* — too many people generating too much waste. This includes sewage, as well as chemical waste generated in the manufacture of consumer products;
- *widespread and ever increasing chemical manufacture and usage, particularly SOCs;*
- *gross mismanagement and irresponsible disposal of chemical wastes* (see chapter 2); and
- *reckless land use practices* that result in runoff of pollutants to waterways.

Despite the goal of the 1972 Clean Water Act to eliminate pollution by 1985, and after spending over $300 billion,[1] surface water pollution remains severe and widespread. In past decades water pollution was highly visible. Laws have reduced the discharge of suspended solids and inorganic waste which gave surface water the "polluted look." Today, concern exists mainly for low levels of colorless and odorless synthetic organic chemials (SOCs) whose long-term (and often unrecognized) effects pose the most significant threat to water ecosystems and water quality.

Surface Water Pollution

There is apparent and unmistakable evidence of the destruction of our coastal waters and marine estuaries. Contamination exists all along the nation's coast — from closed beaches in the east due to sewage and medical waste, to the heavy pollution of San Francisco Bay and the Superfund sites on Puget Sound.[2] Municipal waste treatment plants, factories, and industrial sites continually discharge pollutants from sites near the coasts and from sources along inland tributaries. Runoff from farm land and urban areas, land deposition of air pollutants, and development of coastal wetlands, which can purify water entering them, are all contributing factors. Over 120 million people live within 50 miles of the coast,[3] and development of coastal areas has resulted in increased levels of oil, heavy metals, and other pollutants. Between three and six million tons of oil are dumped into our oceans yearly, mostly due to dumping at sea, petrochemical effluents, and handling spillages.[4]

Surface waters such as rivers can be a special problem as wastewater discharged upstream becomes drinking water downstream. Industries discharge over 18 billion gallons of wastewater daily and 800 million pounds of priority pollutants yearly.[5] Massachusetts Institute of Technology (MIT) scientists identified almost 100 compounds in the Delaware River, a major source of drinking water for many cities and counties. They also noted a high incidence of cancer in people living in the river basin.[6] One study found that people in Ohio who drink surface water have a significantly greater incidence of certain cancers than those who drink groundwater.[7] In 1989, the EPA identified 879 waterway segments in 55 states and territories as contaminated with toxic chemicals.[8] If you add those contaminated by sewage and other pollution the number is over 17,000![9] The U.S. General Accounting Office reports that one-fourth of all dischargers have exceeded limits at least once.[10] This continuously flowing soup makes it possible for industries and water treatment plants to "cheat" on discharge of effluents without being detected. According to one source, over 60% of the industries along the Ohio River do not meet their minimal effluent standards.[11]

Aside from direct discharge by industries into waterways, municipal sources like sewage treatment plants also contribute to toxic chemical contamination of surface water. Municpal plants were generally designed to treat only human and food waste. However, toxic wastes from hundreds of thousands of industrial sources and literally millions of homes are discharged directly into sewers. Although industries discharging to sewers must meet certain pretreatment requirements, in a recent study 41% were found to have exceeded one or more discharge limits.[12] Moreover, an estimated two-thirds of publically-owned sewage treatment plants do not routinely monitor incoming water for organic chemicals or priority metals.[13] In addition, some municipal sewage treatment plants cannot handle the amount of water that comes with heavy rains and are forced to dump raw untreated sewage into surface waterways. In 1984, the EPA estimated that it would take $109 billion by the year 2000 to update wastewater treatment plants "to improve water quality."[14]

Lakes are also a catch-basin for much pollution. Despite our dependence upon them as a resource, the Great Lakes have suffered a long record of abuse. The Lake Erie basin alone contains an enormous industrial base including 17 major metropolitan areas. Water runoff from roughly thirty thousand square miles of the eastern U.S. and Canada drains into the lake. A report by Greenpeace on its 1988 analysis of water sampled throughout the Great Lakes paints a picture of a heavily damaged and declining ecosystem.[15] In a study of just one pollution source, chemical contamination throughout Lake Ontario was traced the Hyde Park waste dump near the Niagara River in New York state.[16] The layers of chemical contamination in lake sediment correspond with the historical use of the Hyde Park dump. Wildlife throughout the Great Lakes region suffers from deformities, tumors, and reproductive abnormalities due to their dependence on a contaminated food supply.[17] The National Research Council and the Royal Society of Canada reported that people living in the Great Lakes area are exposed to a greater number of hazardous chemicals than any other similar group in North America.[18]

Not surprisingly, the heavy reliance on surface water sources account for certain problems in the treatment of drinking water. A study by the National Wildlife Federation found that in fiscal year 1987, nearly 37,000 public drinking water supplies had over 101,000 violations of the Safe Drinking Water Act. These violations affected roughly 37 million people and included exceeding Maximum Contaminant Levels (MCLs) for certain chemicals and noncompliance in reporting and testing.[19] A report by the Center for Responsive Law determined that one in five drinking water systems have detectable levels of unregulated contaminants.[20]

Groundwater Pollution

In the 1980s, concern shifted from surface waterways to groundwater as incidents of contaminated wells became more frequent. Laws passed in the 70s to protect surface waters favored land disposal of wastes. These practices, along with heavy agricultural chemical usage, resulted in increased groundwater contamination, and consequently increases in drinking water well contamination:

- In a study of rural drinking water wells, scientists found that two-thirds exceeded at least one MCL for drinking water. This increases to nearly 80% if lead and turbidity are included. Surprisingly, the west showed a higher occurrence of contamination with over 75% of homes tested exceeding at least one MCL.[21]
- An estimated 3 million people have been affected by well closings in New Jersey, Delaware, California, and New York alone.[22]
- In one study in California, more than one quarter of the wells tested were contaminated by pesticides.[23] A total of fifty-two different pesticides have been detected in California groundwater.[24] An estimated one-fourth of the usable groundwater in the San Joaquin Valley of California, 30 million acre feet, has been contaminated with the pesticide dibromochloropropane (DBCP).[25]
- In a March 1986 study, the Iowa Department of Water, Air, and Waste Management reported SOCs, including pesticides, have

been found in half of the city wells in Iowa.[26] Nitrate levels exceed federal limits in an estimated one-fifth of Iowa wells.[27]

Today reports of drinking water well contamination by pesticides, industrial chemicals, or landfill runoff are becoming so frequent they are no longer media news. Some government environmental officials were often quoted as saying "the more we look, the more we find!"

Regulation of Chemicals in Drinking Water

The Safe Drinking Water Act (SDWA) requires water companies serving more than 25 people to routinely monitor for certain contaminants in the water supply. The SDWA Amendments of 1986 require the EPA to regulate 83 contaminants (Table 3-1). Under the amendments, the EPA is permitted to make substitutions for any seven pollutants on this list. The EPA is planning to add aldicarb sulfoxide, aldicarb sulfone (metabolites of the pesticide aldicarb), ethylbenzene, heptachlor, heptachlor epoxide, styrene, and nitrite to the list.[28] The time frame for the regulations to take effect varies with each group of contaminants, but the final deadline is 1991. Additional chemicals that are required to be monitored by certain water treatment companies are listed in Appendix I.

MCLs and Maximum Contaminant Level Goals (MCLGs) exist for all chemicals regulated in drinking water. Theoretically, a chemical ingested at the MCLG, based on currently available data, will not yield health effects. MCLGs for most SOCs are zero. After determining an MCLG, an MCL is set. MCLs are standards that **must** be met, but are usually set higher than MCLGs. This is due to considerations of the technological feasibility and the cost vs. the benefits of limiting the level of the chemical in water.

Although tremendous progress has been made, the current listing of chemicals monitored in drinking water can be termed deficient considering the more than 66,000 chemicals in commercial production. Many contaminants are not regulated in drinking water, consequently, water can legally be safe and still be severely contaminated to the point of serious health risk. However, it is impossible to set standards for all possible drinking water contaminants. There are simply too many chemicals in commercial use. As a result, drinking water standards will likely be continually lacking full coverage and protection.

Those using private wells are on their own. No laws govern the drinking water quality of wells on private property. In reality, any attempts to directly regulate the water quality of each well would be logistically and economically impractical, not to mention a violation of personal

Table 3-1
Contaminants Regulated in Drinking Water under the SDWA of 1986

Volatile Organic Chemicals

trichloroethylene
tetrachloroethylene
carbon tetrachloride
1,1,1-trichloroethane
1,2-dichloroethane
vinyl chloride
methylene chloride

benzene
chlorobenzene
dichlorobenzene
trichlorobenzene
1,1-dichloroethylene
trans-1,2-dichloroethylene
cis-1,2-dichloroethylene

Microbiology and Turbidity

total coliforms
giardia lamblia

viruses
turbidity

standard plate count
Legionella

Inorganic Chemicals

arsenic
barium
cadmium
chromium
lead
mercury

nitrate
selenium
silver
fluoride
aluminum
antimony

molybdenum
asbestos
sulfate
copper
vanadium
sodium

nickel
zinc
thallium
beryllium
cyanide

Organic Chemicals

endrin
lindane
methoxychlor
toxaphene
2,4-D
2,4,5-TP
aldicarb
chlordane
dalapon
diquat
endothall
glyphosate
carbofuran
alachlor
epichlorohydrin
toluene
adipates
2,3,7,8-TCDD (Dioxin)

1,1,2-trichloroethane
vydate
simazine
PAHs
PCBs
atrazine
phthalates
acrylamide
dibromochloropropane (DBCP)
1,2-dichloropropane
pentachlorophenol
pichloram
dinoseb
ethylene dibromide (EDB)
dibromomethane
xylene
hexachlorocyclopentadiene

Radionuclides

radium 226 and 228
radon
uranium

gross alpha particle activity
beta particle and photon
radioactivity

Source: U.S. EPA, **54 Federal Register,** *22 May 1989, p. 22141.*

rights. While the homeowner on a municipal water system generally relies on the water company to reduce the level of water contaminants, the responsibility for assuring safe water from a private well rests with the well owner.

Drinking Water Contamination and Health Effects

Although it was widely known and visually obvious that surface water sources were becoming heavily polluted, it was not until 1974 that an association was made between SOCs in drinking water and health effects. In that year the Environmental Defense Fund released a report suggesting low levels of suspected chemical carcinogens found by the EPA in drinking water from the Mississippi River were a factor in the increased human cancer mortality rates in New Orleans and Louisianna.[29] Another study found two common SOCs, tetrachloroethylene and carbontetrachloride, both in New Orleans drinking water and in the blood plasma of local residents.[30] New Orleans, which lies at the end of the heavily industrialized Mississippi River basin is the recipient of a significant portion of the effluent discharged to waterways in the eastern United States.

Other studies have also linked specific health effects to drinking water pollution. Incidence of childhood leukemia and perinatal deaths in Woburn, Massachusetts were found to be associated with the use of water from eight municipal wells contaminated with chlorinated organics.[31] An increased prevalence of leukemia was also found in small communities in New Jersey with legacies of groundwater contamination.[32] Although results are preliminary, one study in California found that women who reported drinking no tap water had lower birth defect and miscarriage rates.[33] Several studies have suggested an association between chlorinated drinking water ingestion and a number of digestive tract cancers.[34]

Despite these and many other studies finding an association between drinking water contaminants and specific health effects, the public health significance of the widespread contamination of drinking water is still the subject of considerable debate. To what extent drinking water contamination contributes to the health effects of the chemical age is not conclusively known. And it may never be known for certain. The health effects of water pollutants can be masked by many other sources of chemical exposure from the air, food, and the job site, etc. This makes a decisive determination of cause and effect very difficult. However, it would be reasonable for one to assure that chronic exposure to high levels of certain chemical contaminants in a polluted groundwater supply for example, can have an adverse impact on human health.

In any case, people are becoming increasingly concerned about their health and want to be sure the water they routinely consume is safe. The water supply is essentially the homeowner's only constant source of exposure and one he can do something about. If the source of drinking water is a private well, the quality of water changes little over time. Groundwater comes from one generally constant source and its quality can be more easily assessed and controlled than water from an above-ground source.

The contamination of water and waterways is also better understood by scientists than the contamination of food and air. Contamination is more apparent and more clearly defined in water. Consequently, water pollution is often used as a barometer for the severity of contamination of other sources. Water contamination has also been the focal point of environmental scientific research for decades. As a result, analytical testing methods are generally more advanced for water contaminants than for pollutants in food, soil, or air. Dr. Herman F. Kraybill, a noted researcher on environmental cancer, stated, "...water as an environmental stress system is perhaps better identified, classified, and monitored than any other human exposure."[35]

So, many things are in your favor as you set out to learn about common water pollutants and assess the quality of your drinking water. However, the ways of the world also apply to your water quality situation. You are on your own. You may expect little or no help from local officials, polluting companies, or other perpetrators of contamination. Regardless, it is important for you to realize that if you suspect a problem with your drinking water, solving the problem is often within your power. Any contaminants which may exist are very likely those that are routinely analyzed in labs everywhere and treatment methods currently available are capable of effectively reducing the levels of the vast majority of them.

4. THE LIMITATIONS OF SCIENCE & REGULATION

"The ordinary citizen today assumes that science knows what makes the community clock tick; the scientist is equally sure that he does not."
— Aldo Leopold, 1948

Human-created hazards are an inherent part of our world today. How we deal with the risks associated with these hazards has a direct bearing on the quality of our lives. **Risk Assessment** and **Risk Management** are terms that have come to the forefront of regulatory attempts to confine or contain the risks of modern society. A risk assessment is, ideally, a science-based estimation of the potential dangers of specific chemical hazards. Risk management is the subsequent judgment made in the political arena in determining avenues of protection if, in fact, a risk is determined to exist. This is called the two-stage model for regulating risk — a scientifically based judgment precedes political decision making.[1]

Advances in technology have unquestionably brought about a complex society. A complex society requires complex laws to control and govern its day-to-day processes and activities. Environmental laws and regulations alone are amassed in volumes. Legislators of environmental laws depend on science to provide them with the necessary information to make regulatory decisions and manage risks. While no alternative exists, science and legislation are not necessarily compatible partners.

Considerable uncertainty often surrounds risk assessments. Data is often incomplete, inadequate, or inconsistent. This can result in several interpretations of a specific study by different scientists. As William Ruckelshaus, former EPA administrator, pointed out, "a risk assessment study is like a captured spy: torture it enough and it will say anything."[2]

Confusion often exists between proven factual information and judgments made by a particular study. This leads to reliance by the public on subsequent policies that may have a questionable foundation. At best, risk assessment can only be considered to be partially based on pure scientific reasoning. Consequently, methods of determining safe exposure levels are not without fault. This makes regulatory levels open to debate.

The only reason for the existence of risk assessment is the unrelenting demand by regulators shouldered with the responsibility of protecting the public health. Contrary to popular opinion, however, science does not have all the answers. The social demand for knowledge about the health consequences of exposure to environmental contaminants greatly exceeds the bounds of current scientific knowledge. This most certainly includes the science of toxicology. People want a unequivocal "yes" or "no" answer to questions about the hazards of chemicals, and science is often incapable of a definite answer. Legislative bodies however, require conclusive information in order to carry out decision making. Governments are making important, possibly life and death, policy decisions regarding environmental health, using the limited information that science can provide. Ruckelshaus summed up the problems of risk assessment and public policy by noting that science thrives on the unknown, however, laws require more knowledge or certainty than science is capable of providing.[3]

Uncertainty is reflected in the policies and decisions made today regarding environmental problems, including the regulation of environmental contaminants. Scientific uncertainty along with the prospective costs have deterred action by lawmakers and regulators on many environmental issues. The issues of acid rain and global warming are prime examples. Uncertainty has been used as a license for inaction. Similarly, because science cannot prove a correlation between a contaminant and specific health effects does not mean that it is a safe chemical. Many assume that because a problem cannot be proven under the limited scope of current scientific knowledge, no problem exists. Today, many people in government, industry, and society want unwavering proof that a problem exists before action is taken, which is impossible to do within the confines of science today. Some argue that taking any action on a particular issue given this uncertainty is not economically warranted. However, many others consider it necessary to initiate environmental or health protection policies based on the information which does exist. They see "erring on the side of caution" to be clearly superior to allowing continued chemical exposure without the data base to comprehend the effects.

In addition to depending heavily on the judgements of science, the risk management process also considers social, political, and economic factors in determining whether a particular hazard exhibits an excess level of risk. Approximately 50-thousand deaths and countless crippling injuries occur on the nation's highways yearly, but are generally accepted as part of the cost of transportation in our society. Because all of these deaths and injuries do not occur at the same time, our society does not perceive them as significant. However, when a bus accident claims many lives or a passenger plane crash results in multiple deaths it is seen as a tragedy. Likewise, if the public perceives an unacceptable risk associated with the production, use, and disposal of synthetic chemicals the outcry would result in stricter laws and tighter enforcement.

On the political and economic side, certain levels of chemicals are considered acceptable in industrial effluents, waterways, and in drinking water. A company might go out of business if it were required to emit absolutely no air or water pollution and was not permitted to generate any waste. In addition, many people would likely object to the high cost of treating drinking water to the point where there are no detectable levels of chemicals. So, in addition to attempting to focus the conclusions made by science, managing risks also involves a lot of tradeoffs in other facets of its implementation as well.

Science vs. Science

Science can also be its own worst enemy. Science's ability to create environmental problems is far ahead of its ability to solve them. Environmental protection has not kept pace with damaging technological advances. We have the technologies of nuclear power and countless types of chemical processing, but lack the ability to adequately minimize, or neutralize the wastes generated by them. Instead we use primordial methods like landfills, dumps, and incinerators. All these methods turn one problem into another. Acid rain has no immediate technological solution, nor does carbon dioxide and chlorofluorocarbon build-up in the atmosphere, nor does the staggering problem of hazardous waste which overflows into waterways that we use for drinking water.

Industrial product research and development in the free enterprise system and the rush to market often encourages the use of products before the effects of their use, wastes, and byproducts are even considered. Environmentally persistant and highly toxic pesticides, like DDT, created in the 40s were not banned until the 1970s. Adequate laboratory methods to detect common SOC pollutants in water were not available until the late 1970s, after decades of unmonitored discharge into waterways.

Measuring Risks of Chemical Exposure

Often one hears through the news media of a community's concern about the high number of miscarriages, cancers, birth defects, or other health effects in their area. It is usually reported that a study will be done to determine if an unusually high incidence of the abnormality exists in the area. This is an *epidemiology* study. Most epidemiology studies are case-control studies which compare the health effects of the people in the area in question to an "unexposed" group of people in another area. Ideally, epidemiology studies can give a clear link between a pollutant or pollution source and specific health effects. However, this is rarely the case. Mostly, they can only create hypotheses regarding *possible* chemical causes for certain health effects.

One of the chief drawbacks of epidemiology studies is that they are not easily done under carefully controlled conditions independent of interfering factors or variables, like other chemicals or exposure sources. In any urban area, for example, city dwellers are exposed to a myriad of potential chemical hazards. Diet, age, occupation, potential exposure to enumerable chemicals, and other lifestyle factors all serve to contribute to the manipulation and misinterpretation of the results and the margin of error. Considering the number of sources of contamination to which Americans routinely subject themselves, it is extremely difficult to narrow the cause to one chemical or group of chemicals. Picking out a specific chemical cause, or even a specific source (water for instance) of a disease, may be like looking for a rowboat in the Pacific Ocean.

Several other factors also complicate the accuracy of epidemiology studies. One is the latency period between exposure to chemicals and certain health effects. Cancer, for example, could take 20–40 years between the time of exposure and the effects of the disease, making determination of the cause difficult. In addition, even less is known about the combined effects of a number of chemicals. Exposure to more than one chemical can result in synergistic effects — adverse health effects that are more severe than the effects of each chemical separately. In other words, the whole can be worse than the sum of the parts. No one knows beyond an educated guess the effect of a specific SOC in water let alone the health effects of several present at one time.

Another problem with epidemiology studies is that an enormous number of people afflicted with a specific health effect are needed to draw a statistically accurate conclusion on the possible cause. While it may be apparent that a specific health effect is prevalent in a given area, determining the specific cause may not be possible within the confines of scientific and statistical limits.

Nearly all epidemiological studies are criticized by peer scientists — other epidemiologists — for method inconsistencies. As with any type of research, the results of epidemiology studies can be mishandled or incorrectly interpreted. This is seen in countless examples from the fluoride debate to organic chemical contamination studies. The inability to determine accurate details on exposure is a significant methodology dilema. It is generally difficult to determine the specific levels or doses to which humans were exposed.[4] The effects of low level exposures are the most difficult for epidemiologists to identify. The reporting of a negative correlation may not be seen as significant or valid unless it involves prolonged and heavy exposure.[5] Many have little faith in data generated from epidemiology studies and see them as weak and inherently incapable of giving solid, workable information.[6]

Epidemiology studies cannot predict the effect a chemical will have on humans. They are only able to give information on the effects experienced after people have been exposed, and are most effective at identifying past risks. Epidemiological studies, then, are past exposure assessments.

Despite their shortcomings, epidemiology studies are not without merit. Unlike lab animal studies, epidemiology studies deal directly with human populations. Also, epidemiology studies have generally exhibited a "consistent pattern of association between drinking water and cancer mortality rates at certain sites."[7] Similarly, a large number of studies have found statistically significant correlations that point to smoking as a cause of lung cancer and society has come to accept the scientific evidence. And, as seen in Chapter 2, strong associations exist between certain chemicals and cancer incidence or other health effects. Historically, it has not been an absolute requirement that all facets of a disease or its causes be understood in order to take steps to prevent its occurrence.

It is not surprising that in the face of so much uncertainty the public is distrustful of risk assessment and risk management policies. Many feel that these policies are biased and that the control of hazards is out of their hands. Decision making on risk assessment and management should be done openly and give the public a clear account of the risks as currently perceived. In the absence of this, it is up to the individual to keep an open mind, yet remain wary and critical of studies and regulations that determine personal risk. Be cautious of the interpretations of epidemiology studies. In addition, realize science for what it is — a continuing process with progress reports and updates, with very few real *"breakthroughs"* or definitive and indisputable conclusions.

5. GROUNDWATER

"All rivers run into the sea; yet the sea is not full; unto the place from whence the rivers come, thither they return again."
— Ecclesiastes 1:7

About half of the U.S. population relies on groundwater as a source of drinking water. A third of public water systems and about 96% of rural homes obtain their water from underground sources.[1] Groundwater use, however, is not limited to rural sections of our country. Seventy-five percent of major U.S. cities rely on wells for the majority of their drinking water supply. More than 90% of the residents of Florida, New Mexico, Mississippi, Nebraska, and Idaho rely on underground water sources.

Many people think of groundwater as large subsurface rivers or lakes similar to surface waters which flow at a fairly rapid rate. They remember caves or caverns that they may have seen, with large overhanging rock formations and pristine water. This is not typical of groundwater. Groundwater does flow, but very slowly through compressed sand and gravel deposits. Unlike caves or caverns, there are no ceilings or large air pockets. Groundwater may travel as little as a few inches per year or as much as several feet per day, depending on the region and geology.

Two types of underground waterways, or **aquifers,** are believed to exist. In an **unconfined aquifer,** water drains from the ground surface and recharges the waterway which exists above an impermeable rock layer (Figure 5-1). The ceiling of this aquifer is called the **water table.** The recharge rate is directly dependent on precipitation that occurs locally. Consequently, the water table will go up during rainy periods and down during droughts. A **confined or artesian aquifer** is one that is "confined" between two rock layers impermeable to water. Water in a confined aquifer is under more pressure than an unconfined aquifer where the water table exists at atmospheric pressure. In some cases, when a well is drilled into a confined aquifer, the pressure is great enough for the water to flow unaided to the surface. This is called an **artesian well.** Because the aquifer has an impenetrable ceiling it cannot be recharged directly from the surface. Confined aquifers get their recharge from smaller areas sometimes miles away, called **recharge zones,** where the aquifer contacts the surface without the blockage of confining rock (Figure 5-1). Unlike unconfined aquifers, the rate of well pumping will not significantly affect the recharge rate of a confined aquifer. Some believe there may not be a distinct difference between these two aquifer types. Rather, natural groundwater is really a unified system of water moving through cracks and fissures in geologic formations that does not follow a clear-cut pattern.[2]

Groundwater travels toward, and eventually becomes, surface water. Some surface waters may actually be doorways to groundwater. In fact, during dry periods of low surface stream flow, all the water in a stream may come from the discharge of groundwater. So, it is important to be aware that groundwater, and therefore the water from a well, may not be a separate entity from surface water and related pollution problems. In addition, contaminated groundwater can affect the quality of the surface waters as well. An example is acid mine drainage in coal mining regions.

How Polluted is the Nation's Groundwater?

Determining how much of the groundwater in the United States is contaminated is basically educated guesswork. According to Philip Chen, Chief Hydrologist for the United States Geological Survey (USGS), information regarding the state of our groundwater is simply unavailable.[3] Jay H. Lehr of the National Water Well Association does not believe that groundwater pollution is widespread, but foresees that we will find many more instances of severe groundwater contamination in the future. He estimates that so far we have polluted 200 trillion gallons of groundwater, which constitutes between 1% and 2% of our available groundwater resources.[4] In certain industrial areas of the East, he believes this percentage

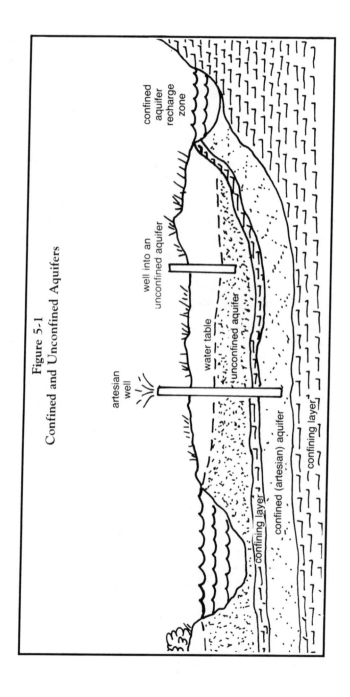

Figure 5-1
Confined and Unconfined Aquifers

could be much higher. The Congressional Office of Technology Assessment (OTA), concludes that contamination is likely to be more widespread and profound than the 1% to 2% estimate indicates, and that most groundwater contamination is not being detected.[5]

According to the USGS, the problem of groundwater contamination is worsening and an increasing amount of groundwater is being contaminated.[6] They list several reasons for this trend:

1. scientists are looking more intently for contamination, and finding it;
2. more groundwater is being developed every year for drinking water;
3. an increasing amount of waste is generated per capita every year;
4. it takes a number of years for pollutants to travel in groundwater, and we are now experiencing the results of mistakes made decades ago; and
5. many people are becoming more aware of groundwater contamination and related environmental problems.

A number of documented cases of groundwater contamination exist across the nation:

- The Long Island counties of Suffolk and Nassau have a population of about 3 million people and rely almost exclusively on groundwater. Yet, over 100 supply wells have been closed because of contamination by synthetic organic chemicals (SOCs), including trichloroethylene (TCE), trichlorethane, and tetrachloroethylene.[7]
- Over one-third of the communities in Massachusetts — a state relying heavily on groundwater — have been affected by chemical contamination.[8]
- The Groundwater Supply Survey (GWSS) indicates that 45% of public water systems using groundwater and serving over 10,000 people had detectable levels of VOCs.[9]

The amount of groundwater that has actually been tested is very small. Where testing of groundwater has taken place, testing procedures have traditionally not been comprehensive enough to give an accurate, overall picture of the water quality, even at specific sites. Also, a large number of substances that can contaminate groundwater are in widespread use and haphazard disposal of the hazardous chemicals has been the rule, rather than the exception. Consequently, more detection of contamination is expected. According to an EPA report to Congress on waste disposal practices, "There are potentially millions of sources of contamination and isolated bodies of groundwater contamination nationwide."[10]

The lack of comprehensive monitoring means most incidents of groundwater contamination are only discovered by accident when a home-owner complains to state or local authorities or when people begin to exhibit health effects. According to the EPA report to Congress, "Almost every known instance of groundwater contamination has been discovered only after a drinking water source has been affected."[11] Response to citizen complaints and routine monitoring account for most of the testing that has been done, but this is not considered a scientific assessment of overall groundwater quality. Historically, large areas of groundwater have been polluted by the time contamination is discovered.[12]

Due to groundwater's characteristically slow movement, pollutants may not be discovered in drinking water wells for decades after their release into the environment. Considering the continued and dramatic rise in organic chemical production, usage, and disposal since the 1930s, coupled with the fact that it could take decades before discovery of the pollutants, groundwater contamination is likely to continue at an even more rapid rate. Even reducing new contamination sources by promptly cutting hazard-ous waste production and practicing proper disposal would not result in an immediate reduction of the incidence of groundwater contamination.[13] Many more of our past mistakes have yet to catch up with us.

Documented instances of groundwater pollution have turned up some 200 different compounds. Many believe this could be a low representation of the actual number of contaminants in groundwater because of the limited amount of testing, the astronomical number of possible contamin-ants, and the extraordinary amount of chemical usage and waste in this country. The scope of testing done at specific sites often overlooks many possible contaminants. In the past, water was often tested only for metals or other inorganics, not SOCs. Only recently have groundwater monitoring efforts begun to include organics, though the total number of SOCs tested for at a specific site is limited by cost and technological capability. Econom-ically it is out of reach to consider comprehensive monitoring of ground-water supplies, or even to initiate research to understand all the possible characteristics of underground hydrogeology.

The OTA concludes that we will very likely never know with any degree of accuracy, the extent of groundwater contamination.[14] The infor-mation available is limited and defined by the scope of specific investiga-tions, i.e., what chemicals are being tested for and in what areas. Deter-mining the extent of groundwater contamination at just one site is a monumental task in itself. A series of monitoring wells must be drilled and extensive laboratory testing must be done to determine the boundaries of the contamination. It is very difficult for scientists to be sure that samples are taken in the correct locations, at the right depth, and the tests are run properly, to locate all contaminants and outline the dimensions of the plume. To determine the potential movement of the pollutants and whether it will affect drinking water supplies, hydrogeologists must determine:

1. the direction the aquifer is flowing;
2. the dimensions (width, length, and depth) of the contamination; and
3. the geologic makeup of the aquifer, which determines the flow and speed of the aquifer and the pollutants.

Though no one really knows how much of the nation's available groundwater is contaminated, the OTA points out the effects of even a small percentage of the total being contaminated are potentially very significant.[15] Often contaminated groundwater is found in industrial or heavily populated areas. Wherever people are living and working, the potential for groundwater contamination most certainly exists. Humans, their waste, and their water supplies are always found together. If the groundwater supply happens to be the only source of potable water in the region, then obviously contamination would have a definite impact on the area in terms of health, industry, and the economy.

Groundwater is not nationally protected, although provisions for identification and protection of sole-source aquifers were included in the Safe Drinking Water Act (SDWA) amendments of 1986. Some argue that comprehensive legislation on groundwater would be difficult considering its lack of uniformity and geographic diversification throughout the U.S. with respect to geology and hydrology. Others argue that because groundwater does not follow state boundaries, protection would not be uniform under state law and adjoining states with lax provisions could adversely affect the groundwater of a state with strict protection provisions. The EPA has urged each state to set its own laws, as hydrogeological aspects of groundwater within states would be more uniform. Groundwater protection strategies have also been issued by the EPA to help states initiate their own protection programs.

Movement of Groundwater Contaminants

Like surface water, groundwater moves at the mercy of gravity. Consequently groundwater movement is essentially downhill, and the water

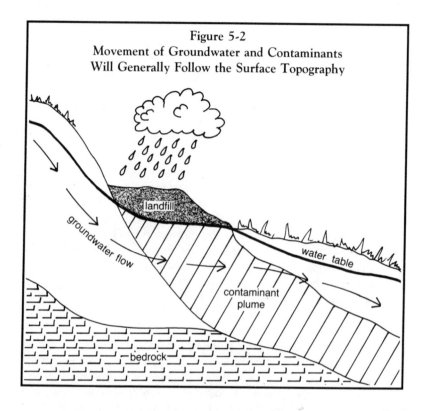

Figure 5-2
Movement of Groundwater and Contaminants
Will Generally Follow the Surface Topography

table generally follows surface topography (Figure 5-2). Water supply wells, then, should not be placed at a lower elevation than septic tanks, waste disposal sites, and landfills. Since groundwater generally flows very slowly from a higher to a lower level, a well located several hundred yards down from a pollution source may not be affected for decades.

Water moving down through soil toward an aquifer is usually cleansed as soil and naturally occurring microorganisms filter out or break down foreign matter and contaminants. However, SOCs, and in some cases excess amounts of other types of waste, override these natural breakdown and filtering processes. Microorganisms in the soil cannot break down SOCs which are foreign to them, and the soil cannot accommodate a high amount of material leaching through it. Once a contaminant reaches the aquifer, very little, if any, further cleansing takes place because aquifers are chemically reducing (lack chemical activity), cool, dark, and lack oxygen and microorganisms.

The type of soil overlaying unconfined aquifers is a factor in the likelihood of groundwater and well contamination. Clay does not have the drainage capability that sandy soil does. Therefore, water and contaminants will travel through it more slowly. Knowing soil type and well depth, one may be able to roughly judge the likelihood of contamination of the well. A shallow well in sandy soil has a relatively good chance of contamination. On the other hand, if the soil in the area is non-porous and clay-like and the well is deep, the chance of contamination is less likely.

In some areas, fractures or macropores in the Earth's surface are faster routes to groundwater for contaminants than filtration through soil layers. These channels in bedrock or soil layers can serve as contaminant pipelines to the water table. This could cause some compounds that are normally biodegraded or held by the soil to reach groundwater. Waterborne diseases have been reported to increase in the spring and the fall. This may be due to the rapid recharge of aquifer water through these fractures.[16] No natural purification occurs under these conditions.

Incorrectly built wells or improper drainage around the well can be a factor in direct contamination of the water source. Polluted surface runoff can drain into the well if the well-casing is not sealed properly and the ground surface is not sloped away from the well.

Movement of groundwater pollutants within an aquifer is complex and not well defined. Pollutants disperse or diffuse and spread out as they move within an aquifer. It is very difficult to predict dispersion and movement of specific contaminants in groundwater. Only extensive testing and field study will determine the spread of contamination at a specific site with any degree of certainty.

Pollutants in groundwater move in a **plume** — a concentration or clump of the contaminant(s) within the aquifer. The plume will generally be carried with the flow of the water, spreading out and down, mimicking the level of the water in the aquifer. The levels of pollutants will generally be lowest on the fringes of the plume and the highest concentrations will likely be closest to the source.[17] Contaminants may move slower, faster, or at the same rate as the water, depending on the specific chemicals involved. Contaminants that dissolve well in water usually flow along with the water in the aquifer. Volatile organics which are lighter — less dense — than water will usually float near the surface of the aquifer. Volatile chemicals that are insoluble in water, like hydrocarbons from petroleum refining, are a good example. It is believed that the surface tension between the hydrocarbons and the aquifer material retards the movement of the hydrocarbons with respect to the groundwater flow.[18]

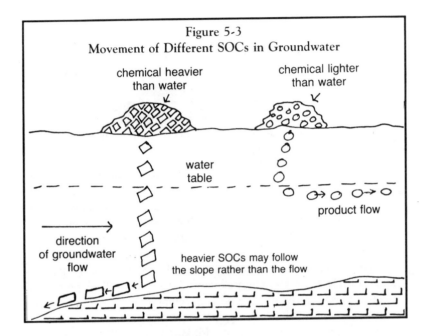

Figure 5-3
Movement of Different SOCs in Groundwater

Some heavier halogented organics have a lower viscosity and higher density than water and may not follow the flow of groundwater as readily (Figure 5-3).[19] While a portion may dissolve in, and move with, the groundwater, most of these contaminants will likely settle to the bottom of the aquifer. Reactions between different chemicals and the geologic structure of the aquifer can also affect the movement of the plume.[20]

Most of the information available on plume movement is based on computer models from limited laboratory data. These models are controversial and disputable in their practical application to "real-life" cases.[21] In any case, it is known that contaminants in groundwater can travel in three ways:[22]

1. In separate or individual "islands" of contamination within the water of the aquifer. Periodic filling and overflow or seepage from waste sites or impoundments, and leaching of contaminants — as would be the case after a heavy rain — will result in "groups" of contaminated water moving with the aquifer flow. This is an example of why periodic analysis of water is in order in some cases. The contamination may not be discovered if only one analysis is performed.

2. In an increasing plume caused by continuous movement of contamination into the aquifer. If a number of pollutants are present, the plume may actually be a consecutive group of smaller plumes of individual contaminants. In this case, a single sampling and analysis may not identify all pollutants.
3. Large widespread contamination from no clearly defined point source, as would be caused by large numbers of septic tanks or widespread pesticide spraying. Contamination of this type would encompass a large area without easily defined boundaries.

Overpumping a well can also be a factor in pollutant movement in groundwater. In the area immediately around a well, the water level of the aquifer will decline slightly. This is a natural phenonmenon based on the same principle as a vacuum. Water drawn out of a well creates a void that must be filled. This is done by pulling in water from the area around the well creating what is called a **cone (or zone) of depression** (Figure 5-4). This cone, or pulling force, may extend for miles or only a hundred feet.[23] This force can also pull pollutants toward the well. A plume may be pulled toward a well more rapidly by pumping.

Low water levels due to overpumping can cause unwanted mineralized water, called brine, to infiltrate the drinking water well.[24] The increased pressure that occurs during pumping can cause flow between two different aquifers. Mineralized water, usually high in sodium or other inorganics, from a deeper aquifer can be pulled through confining layers between the aquifers, increasing the salt content of the upper aquifer, possibly to the point of contamination (Figure 5-4). In areas near the coast, sea water might be pulled into the aquifer. This movement of highly mineralized water from lower adjacent aquifers usually affects deeper wells rather than shallower ones.

Chemicals in Groundwater

Groundwater can contain a wide variety of natural and manmade contaminants including: inorganics, like nitrates and chlorides; radioactive elements; heavy metals; and SOCs. Chlorinated volatile organic chemicals (VOCs) are produced and used by industry in large amounts and are very mobile in groundwater. VOCs seem to travel easily through overlaying soil and gain rapid and relatively unhindered access to groundwater systems. Consequently, they are considered to be prime candidates for groundwater contamination. In the Groundwater Supply Survey (GWSS), which tested for 29 VOCs in nearly 1,000 groundwater sources, the chemicals most

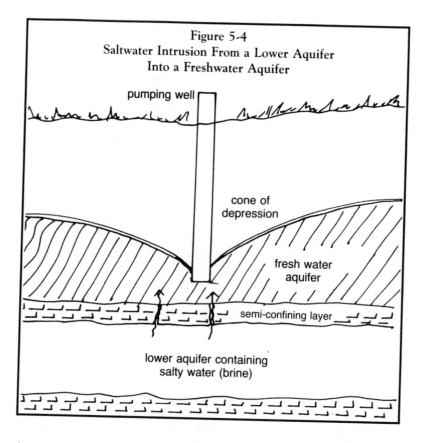

Figure 5-4
Saltwater Intrusion From a Lower Aquifer
Into a Freshwater Aquifer

pumping well

cone of
depression

fresh water
aquifer

semi-confining layer

lower aquifer containing
salty water (brine)

frequently detected were tetrachloroethylene, 1,1,1-trichloroethane, trichloroethylene (TCE), 1,2-dichloroethylene, and 1,1-dichloroethane.[25] Spills of even small amounts of organic chemicals can contaminate large amounts of groundwater. Five gallons of perchlorethylene (PCE) spilled on the ground, for example, will likely contaminate 50,000 gallons of groundwater to unhealthy levels.[26]

While testing is the only way to obtain an accurate picture of water quality a well, general conclusions can be made about the occurrence of SOCs in groundwater to aid in test selection.

1. chlorinated VOCs like TCE and 1,1,1-trichloroethane have frequently been found in groundwater;
2. public drinking water systems using groundwater frequently have
   VOC contamination; and

3. usually when a supply is contaminated by SOCs, two or more are found in the water supply.

Heavier organic chemicals do not pass as easily through soil barriers. This does not mean that they have not been found in groundwater. They have. In a survey in New York state, for example, three members of a class of heavier chemicals called phthalates were among the ten most frequently detected compounds.[27] In some cases, testing only involved chlorinated SOCs. So, it is possible that in many instances other non-volatile chemicals may have been overlooked in the testing and some surveys. The OTA in its study on groundwater was unable to determine whether past surveys had not analyzed for SOCs, or whether they had not been detected in the testing.[28]

As a general rule of thumb, surface waters are more polluted than groundwaters. This is due to disinfection byproducts and the heavy industrialization of river and lake basins. However, when groundwater is contaminated, it generally involves higher concentrations of chemicals due to the low amount of dispersion, mixing, and dilution. As an example, the highest level of TCE reported in surface water was 160 ppb, but 510,000 ppb was the highest level found in groundwater.[29]

Surface water quality, unlike groundwater, will fluctuate and vary dramatically with time. Surface water is faster moving, and mixes much more rapidly than does groundwater, making it harder to pin down contamination. The water in a well generally will not change dramatically over a short period of time — say a year. The level of a contaminant may change as a large plume moves into the area of the well, but if the well proves to be uncontaminated after comprehensive testing and no potential sources of contamination exist nearby, then the quality is unlikely to change any appreciable amount in the near future. Generally speaking then, it is better to have groundwater as a drinking water source unless, of course, it is found to be contaminated.

In most cases of groundwater contamination, only the higher aquifer is affected, though this is not always the case. Sampling of aquifers at different depths in New Jersey found VOCs in 10% of the wells tested in the upper aquifer, 22% of those tested in the middle one, and 28% of those in the lowest.[30] This apparent heavier contamination of the lowest aquifer was thought to be the result of leakage from the middle aquifer and the large number of wells bored into the lowest aquifer in overdeveloped areas.

Compounds found in surface water are often different than those found in groundwater. VOCs have traditionally been found in contaminated groundwater, but are not found at elevated levels in unchlorinated surface water because they volatilize easily into the air. Chemicals generally break down more easily in surface waters than groundwater because more microorganisms exist in surface waters to act on them. For example, the second-generation pesticides in use today degrade rather quickly in surface water. They also biodegrade in soil. But, if they should reach groundwater, where there is usually an absence of microorganisms to degrade them, these pesticides could persist for some time. Consequently, chemicals that would not ordinarily persist in the surface environment, will remain for a longer time in groundwater.

Generally, if one uses groundwater from a private well and it is contaminated, he will be able to find the chemicals responsible if enough testing is done. In the case of surface water, the extraordinary number of possible sources of chemicals, the rapidly changing status of overall water quality, and the additional number of chemicals created by chlorination, makes determining all possible contaminants in the original source water nearly impossible.

Cleaning Up Contaminated Groundwater

Removal or cleanup of a large contaminating source will not make an aquifer clean. Crude and primitive methods like building clay dams or digging trenches to contain the spread of contamination are limited in their effectiveness and usually work only when the contamination is from a single source and the contaminated area is small.[31]

Ways of treating large quantities of contaminated groundwater do exist, mainly on the municipal scale. Groundwater treatment or reclamation involves the pumping of the contaminated water to the surface to treat it. In some cases, the water is then returned to the subsurface. Activated carbon and air stripping, which move the pollutants from the water into the air, are examples of SOC removal technologies. The cost to treat the immense volumes of water required to clean an aquifer are nearly unimaginable, and some experts believe it is nearly impossible to effectively treat many cases of organically contaminated groundwater to a drinkable level.

The cost to reclaim only one groundwater pollution site is staggering. The bill for cleanup of contamination at the Rocky Mountain Arsenal in Colorado, where the military began dumping pesticides and other organics during WWII, is expected to be almost $2 billion.[32] Merely determining the size of a plume costs hundreds of thousands of dollars. Accurate determination of aquifer contamination requires comprehensive monitoring and the analysis of enormous numbers of lab samples for many possible pollutants. It is also possible that not all pollutants may be identified. Needless to say it would be useless to treat the water without removing all pollutants. The construction of a treatment operation could cost well into the millions of dollars, and operating costs for a project that will probably take decades to complete will not be pocket change either. Also, costs for these types of projects increase substantially every year. In addition, the public's perception of using drinking water that was once measurably contaminated may be less than favorable, even if it meets safety standards.

Where large areas are contaminated and a specific source cannot be pinpointed, as can be the case with agricultural application of pesticides, the treatment of groundwater is unrealistic. The use of the pesticide dibromochloropropane (DBCP) in the San Joaquin Valley of California between 1955 and 1979 encompassed about three million acre feet of surface land and 20 million acre feet of groundwater.[33] The dilemma of simply defining the boundaries of such a huge contaminant plume would be a monumental task.

With clean up of contaminated groundwater neither technologically or economically practical, all that remains is prevention. Government officials and groundwater experts in the field agree that when it comes to groundwater contamination, the best solution is prevention, and preserving the resource is of the highest importance.

6. WATER HARDNESS & Cardiovascular Disease

"The water pure that bids the thirsty live."
— Ellen H. Underwood (1845–1930)

Hardness is not a water contaminant, although many people treat their water as if it were. Rather, it is a characteristic of water. Hardness is determined by the calcium and magnesium content of the water supply. Since calcium is the major constituent determining the degree of hardness, it is used as the principle test constituent. Hard water is loosely defined as water having more than 75 ppm of calcium carbonate (Table 6-1). The extent of water hardness tends to vary with geographic area (Figure 6-1).

Concern about the level of hardness in water is principally an economic one. First, hard water requires the use of more soap than soft water because the calcium and magnesium "tie-up" the soap, reducing its ability to lather, and consequently its cleansing power. Secondly, calcium and magnesium come out of liquid solution to form a solid precipitate called scale. This can clog hot water pipes, water heaters, and boilers. These problems are purely economic, and water softener company marketing strategies are based heavily on the amount of money one might save

Table 6-1
Degrees of Water Hardness

Concentration of Calcium Carbonate (mg/L)	Extent of Hardness
0–75	soft
75–150	moderately hard
150–300	hard
over 300	very hard

by purchasing a softener. However, it is advisable to do some cost comparing before investing in a softener which could run $1,200 or more. In addition to the cost of the softener, one must regularly purchase salt to keep it in working order. Water softeners used in homes are usually the ion-exchange type. They utilize a resin to exchange sodium for the calcium and magnesium in the water (Figure 6-2).

The primary purpose of this book is to increase awareness of drinking water as a source of exposure to pollutants and act as a guide in assessing the quality of the water source routinely used. However, continually mounting evidence is pointing to some minerals associated with hardness as having some sort of preventive effect against the development of cardiovascular disease. Consequently, the removal of these minerals may be a factor in the occurrence of the disease.

A study in 1960 was the first in the United States to implicate the degree of water hardness as being a factor in the development of degenerative cardiovascular disease.[1] Since then, numerous studies suggesting a link between water hardness and human health have been published, principally in the United States, the United Kingdom, and Canada. The results show a clear and statistically significant relationship between the level of hardness in drinking water and heart disease, hypertension, and stroke.[2] These studies have generally supported the contention that the softer the water the higher the incidence of cardiovascular disease. According to the National Academy of Sciences (NAS), studies in the U.S. and Canada reflect cardiovascular mortality rates 15 – 20% higher in populations with very soft water as compared to those with hard water. In the U.K. the cardiovascular mortality rates are closer to 40% higher.[3] These are rather surprising findings when one considers cardiovascular disease is the leading cause of death in the U.S., accounting for a full half of all

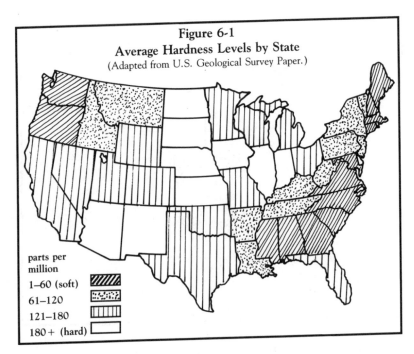

Figure 6-1
Average Hardness Levels by State
(Adapted from U.S. Geological Survey Paper.)

parts per million
1–60 (soft)
61–120
121–180
180 + (hard)

deaths. Although the absence of certain water components is not the sole consideration in the development of cardiovascular disease, it is likely to be a serious potential hazard to people with arteriosclerosis.[4]

The relationship between water and cardiovascular disease is further supported by the fact that cardiovascular disease rates measurably vary by geographic area.[5] Water components like those associated with hardness also generally follow a geographic pattern.

Scientific evidence suggests that certain trace elements, common to heavily mineralized hard water, are required for numerous biological processes which take place within the human body. Enzyme reactions and the selective permeability of cell membranes are two such processes. Cardiac cells may also be at the mercy of the mineral balance existing within the body. When water has an imbalance of these minerals, the heart's function may be affected.[6]

The specific reason for these suspected adverse health effects from soft water is not known, but generally three theories exist:[7]

1) *Calcium and magnesium in hard water act as protective elements, guarding against the health effects from certain contaminants.*

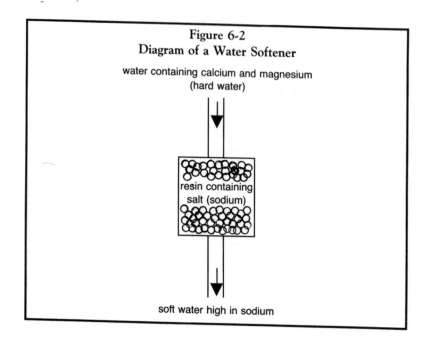

Figure 6-2
Diagram of a Water Softener

water containing calcium and magnesium
(hard water)

resin containing
salt (sodium)

soft water high in sodium

Researchers have found lower levels of magnesium in certain tissues (including the heart) of people who died due to myocardial infarction than those who died in accidents.[8] Some hypothesize that calcium and magnesium, which are easily taken up by the small intestine, overshadow and reduce the movement and effects of lead within the human body.[9] This, along with the damaging effects of the sodium added to the water by softeners is the basis for the majority of studies on the health effects of hardness in drinking water.

2) *Other trace mineral elements present in hard water (and removed when water is softened) have a protective effect.* Vanadium, for example, a trace element, may have essential nutritional aspects and reduce serum cholesterol levels.[10]

3) *Soft water corrodes toxic metals like lead from plumbing, solder, and water distribution systems.* Heavy metals like lead are associated with cardiovascular disease and are found at higher levels in water flowing from taps in soft water areas. Limited information also suggests the presence of cadmium, copper, and zinc may be associated with initiation of cardiovascular disease. All of these metals are found in plumbing materials and have been proven to dissolve into soft drinking water. Generally,

hard water is not corrosive.[11]

The NAS does not believe enough information exists to alter the level of hardness or softness of water supplies on a national basis. But it cautions waters companies softening their water to carefully evaluate the need for it.[12] If a direct cause/effect relationship is proven between cardiovascular disease and drinking water hardness, the NAS estimates that mortality from cardiovascular disease could be measurably lowered in the U.S. each year by appropriate drinking water treatment. When one takes into account that as many as 15% of the one million avoidable cardiovascular deaths annually may be associated with soft water, it certainly puts things into perspective.[13]

Until the jury is in, the smart money is on those who do not soften their drinking water. Those that live in hard water areas and are concerned about how water hardness could reduce the lifetime of the water heater and other appliances, or want to save a little on their soap budget, can soften only the hot water.

7. HEAVY METAL POLLUTANTS

"Water is much more wholesome from earthenware than from lead pipes . . . water should not be brought in by lead pipes if we desire to have it wholesome." [1]

— Vitruvius, Roman architect and engineer

Metals are elements (remember the periodic table in high school chemistry?) that naturally occur in mineral ores found throughout the earth's crust. They would have remained there if man had not stirred them from these natural mineral states for use in industrial processes, fabrications, or chemical reactions. During mining and industrial processes a great deal of waste containing heavy metals is generated, including mine tailings, waste piles, and process wastewater effluent from factories or mills.

Metals removed from their natural ore state become more heavily concentrated, resistant to breakdown, and are not easily accepted back into the chemical cycles of the natural world. Their residence time as a pollutant before acceptance back into the ecosystem could be decades or even centuries. The form a metal takes becomes altered from its natural state and in some cases to forms more readily absorbed by humans through the intestines.

Heavy metals, simply by virtue of their weight, will generally deposit in the sediment of surface waters. However, stirring up the sediment by heavy stream flow or dredging will release the metals and cause elevated levels in the water. In addition, deforestation, mechanical disruption of the ground surface, and acidic precipitation cause metals to leach from soil into waterways.

Health Effects of Metals in Drinking Water

Toxicity levels vary widely among different metals. Certain metals exhibit more toxic effects than others and therefore are of more concern. Calcium, zinc, and iron, for example, are beneficial metals, but others like lead, cadmium, and mercury are all highly toxic. Arsenic, mercury, lead, cadmium, chromium, and copper are all regulated under the Safe Drinking Water Act (SDWA). Specific health effects of each are discussed in the following sections.

Sampling Water for Metals Testing

Polyethylene bottles with a polypropylene cap are used in the sampling of all metals. A preservative, nitric acid, is used to contain the metal and minimize interference in the test. Obtain a sample bottle with the preservative from the laboratory. If samples must be mailed to the lab, do not request a nitric acid preservative as it is illegal to send through the mail. Simply request a styrofoam mailing container and cold packs. Refrigeration is a suitable preservative until nitric acid can be added to the samples in the laboratory.

Because the plumbing system is the greatest source of most metals, the sampling procedure must isolate the different components of the water distribution system to determine where a problem may exist. It is best to take four samples. Do not use any water after 11 p.m. the night before samples are taken and follow this procedure:

1. In the morning before any water is used (this includes toilet flushing), open the cold water faucet and immediately fill the first bottle to one inch below the top. Do not splash the preservative out of the bottle. Turn off the water and cap the bottle. Label it *faucet.* Ideally this sample bottle should hold only 100 milliliters (about 3 oz.) to be sure only the faucet and its connectors are being sampled.

2. Turn the water on. Wait a few seconds, then fill the second bottle and recap it. Label it *interior plumbing.* This bottle should hold a liter (about a quart) of water to obtain a representative sample of the indoor plumbing.

3. Turn the cold water faucet back on. Wait until the water turns cold. Immediately fill the third bottle and recap it. Label it *service line* or *exterior plumbing* for private well sources.

4. Let the water run for another three minutes and fill the fourth bottle. Recap the bottle. Label it **water main** or **well.**

Use this method in sampling for all metals. Follow these instructions exactly. Sampling is as critical as the analysis in obtaining accurate results.

Testing for Metals

Metals analysis is less complex and relatively inexpensive compared to tests for synthetic organic chemicals (SOCs). All metals analysis can be done on the same instrument and the procedure is nearly the same for all, except mercury. Often a lab will charge a flat rate for the first metal, then somewhat less for each additional metal tested. If a metal not discussed in this chapter, such as aluminum, sodium, or another metal found in Table 7-1, is of concern and one is interested in having their water tested for it, it is very easy to add it on to the request. When communicating with the laboratory, request **metals analysis** and indicate which metals are to be tested.

All pollutant concentrations may change over time. Water samples tested for metals are no exception. One result should not be used to make decisions. If the metal is found in plumbing materials (Table 7-1), take four samples as discussed above, evaluate the results and retest as a check of the accuracy of the results.

LEAD

Due to its low cost and high durability, lead was initially used as distribution material for drinking water systems in the United States. Today, approximately 51% of U.S. cities still use lead or lead-lined pipes. Many of these cities, like Boston, still have some of their original plumbing. An estimated 71,000 homes and apartment buildings in Washington, D.C. get their tap water through lead pipes.[2]

The 1986 amendments to the SDWA require the use of lead-free (less than 8% lead) pipe, solder, and flux in the installation or repair of public water systems and interior plumbing in homes and buildings. However, the use of lead solder was very widespread until the recent ban, and existing plumbing is not affected by the amendment.

Sources of Lead in Drinking Water

Although lead has been mined and processed for thousands of years, its mass accumulation and distribution throughout the environment increased significantly with industrialization. Studies of ice layers in Greenland found significant increases in lead concentrations in the layers of of

Metal	Chemical Symbol	Priority Pollutant	Regulated or Considered for Regulation in Drinking Water	Known to be Used in Plumbing Material
Aluminum	Al			
Antimony	Sb	X		X
Arsenic	As	X	X	
Barium	Ba		X	
Beryllium	Be	X	X	
Cadmium	Cd	X	X	X
Chromium	Cr	X	X	
Copper	Cu	X	X	X
Lead	Pb	X	X	X
Mercury	Hg	X	X	
Nickel	Ni	X	X	
Selenium	Se	X	X	
Silver	Ag	X	X	
Sodium	Na		X	
Thallium	Tl	X		
Vanadium	V		X	
Zinc	Zn	X		

Table 7-1
Common Metals

ice corresponding to the years after 1750 — the dawn of the industrial revolution. A sharp rise also appears after 1940 due to the post-war industrial boom. In 1965 the lead level in the snows of Greenland was 400 times the natural level in 800 B.C.[3]

Approximately 65% of the lead currently mined is used in storage battery manufacturing and production. Ten percent is used as gasoline additives for pre-1975 automobiles. Other sources of lead are shown in Table 7-2. The main source of lead in drinking water, however, is the components of water distribution systems and the plumbing systems in homes and buildings.[4] This accounts for a significant amount of the total lead exposure to humans.

The adverse health effects of lead in plumbing systems have long

---

Table 7-2
Sources of Lead[5]

- petroleum and gasoline additives for pre-1975 autos
- solder for joints in food cans and water pipes
- sheet and pipes in the building industry
- solder and bearings for products like automobile engines and washing machines

- storage batteries
- paints
- radiation shields
- plastics
- ceramics
- makeup

---

been ignored. Only recently have regulations to control lead in drinking water been implemented. Lead was recognized as a poison as far back as the Roman Empire. Vitruvius, a Roman architect and engineer, even postulated that lead pipes used to transport water could contaminate the water and therefore the body.[6] In 1845, warnings in Boston of the dangers of lead in drinking water were ignored. A report by the Boston Commissioners stated:

> "Considering the deadly nature of lead poison and the fact that so many natural waters dissolve this metal, it is certainly the cause of safety to avoid, as far as possible, the use of lead pipe for carrying water which is to be used for drinking."[7]

In 1871 a Massachusetts Institute of Technology professor reported the dangers of lead pipe usage for drinking water. He determined that water standing in pipes will contain higher amounts of lead than water flowing and that temperature will affect the rate of pipe corrosion. He also cited several cases of lead poisoning from drinking water.[8]

Wellington Donaldson documented the extent of poisonings in 1924. He wrote:

> "Obscure symptoms, lack of thorough investigations, and fear of arousing public apprehension or of inviting litigation are the principal causes for silence or complacency as to possible plumbism, on the part of cities having a large proportion of lead services."[9]

It was not until 1982 that the EPA reported that the largest source of lead in drinking water is due to the corrosion of plumbing materials.[10] These materials include:
- water mains made of lead (rare);
- lead gooseneck or pigtails — short sections (usually six to eight

feet long) of pipe connecting the water main to the service line or the service line to the meter;
- service lines and interior plumbing made of lead;
- lead solder and flux used to connect copper pipes; and
- faucet fixtures made with lead alloys, like brass or bronze. [11]

While any water is corrosive to some extent and can leach some lead from plumbing materials, **lead solvency** — how much lead will dissolve from the plumbing materials into the water — is determined by certain properties of the water itself. These properties include the alkalinity, the pH, and the softness of the water. Very corrosive waters are soft (less than 60 ppm as calcium carbonate), have a low pH (less than 8.0), and are low in dissolved solids and alkalinity (less than 30 ppm). [12]

So, the softer and more acidic the water, the more corrosion occurring in the pipes, and the more lead in the water. For example, if the pH is 6.3, the hardness is 3.8 ppm, and the alkalinity is 4.0 ppm, the water is highly corrosive. Why? Because the pH is low (below 7), the hardness is far below the upper limit ceiling of 60–75 ppm which classifies it as soft water, and the alkalinity is too low to neutralize the acidic properties of the water. Keep in mind that this example is the "worst case." The degree of corrosiveness of the water will vary depending on its properties. It may have no corrosive properties at all, however, the pipes will still leach some lead simply due to water and pipe contact. It will be minimal when compared to the "worst case," however, the only way to know the level of lead in the water is to test it. As a point of reference, about half of the 100 largest cities in the U.S. have corrosive water. [13]

Studies show that roughly 1/3 of the country has very soft water (less than 60 ppm calcium carbonate), and 80% of public water systems have moderately to highly corrosive water (pH of 8.0 or less and/or an alkalinity of less than 30 ppm). [14] Groundwater in New Jersey's coastal region was reported to be acidic enough to cause high lead levels in the tap water of certain homes. In some areas of New Jersey, lead in the tap water was over 100 ppb, while the level of lead in the groundwater itself before entry into the house was 10 ppb. [15]

Municipal water companies with water classified as corrosive will be required to practice some kind of corrosion control, if they do not already. This typically consists of the addition of a neutralizing compound like lime or sodium hydroxide to lower the the pH, or the addition of corrosion inhibitors. Corrosivity of water varies from state to state and, particularly in the case of rural water supplies, from well to well. A water company may not feel that its water is particularly corrosive but its analysis of lead is based on water tested at the plant, not water that has been through the distribution system and sat in the pipes overnight.

Temperature is also a factor in corrosion. Hot water substantially increases the corrosion capability of the water, and therefore the amount of lead leaching from pipes. Studies have found high lead concentrations in samples taken from the hot water tap. Scottish researchers found lead levels in hot water up to 14 times higher than that of the cold water.[16] So, it is very likely that the hot water levels could be quite high, even if cold water levels are very low. In addition to the potential for corrosion of lead pipes or solder in the hot water lines, it is possible that the hot water heater is lead-lined. Some people, thinking they are cutting corners, will draw water from the hot water tap prior to heating it for cooking. Using water from the hot tap for consumption is by no means a healthful practice due to the leaching of several metals in addition to the presence of a variety of microorganisms.

The insertion of a section of copper pipe to repair lead pipe can also cause plumbing corrosion. This method of repair could result in even higher lead levels in the water than if the plumbing remained entirely lead. The chemical properties of both copper and lead result in an electrochemical reaction across the junction or seam of the repair causing corrosion of both metals into the water. The repair process also may cause chipped copper particles to deposit on the walls of the lead pipe creating more electrochemical reactions. In addition, lead particles chipped off in the installation can also cause an increase in lead levels, particularly if they end up in the strainer at the faucet. It is a good idea to check the strainer and dump particles regularly, especially after plumbing work.

Corrosion can be enhanced by another electrical-based corrosion source — the practice of grounding electrical equipment and appliances to water pipes. However, they should not be removed unless a qualified electrician installs an acceptable alternative.

About 240 million people in this country drink water from soldered pipes,[17] and chances are, due to the lower cost of lead-based solder compared with tin-silver or tin-antimony, the great majority of solder found on pipes today is lead. Use of lead solder in newer homes may be a greater concern than existing lead pipes in older homes. In hard water areas, calcium carbonate scale can form on the inside of pipes, insulating the metal from the water and reducing lead solvency. This process takes place over time, consequently, older homes with hard water are more likely to have this protective coating on the inside of the pipes. In new homes with lead solder joints, lead water concentrations are likely to be very high initially due to the lack of any insulating coating and the dislodging of solder particles from the recent plumbing work. The Nassau County, New York Health Department tested new homes and found levels as high as 17,000 ppb at a neutral pH of 7.[18] A Minnesota multifamily building site had a level of 33,000 ppb.[19] New construction had been mainly 50–50 tin-lead solder until the recent ban. The primary alternative solder being used after the ban is 95–5 tin-antimony.

While pipes and solder may not be the only sources of lead in water, other sources could be negligible by comparison. Those with new lead solder in their plumbing and served with corrosive water would have drinking water as their primary source of lead exposure.[20] People living in homes less than five years old with lead solder are at the greatest risk of high lead levels in drinking water.[21]

A major source of lead in surface water is atmospheric particulates, generally from leaded gasoline combustion, ore smelting, and the burning of fossil-fuels. Atmospheric emissions declined dramatically in 1985 according to the EPA. A 48% reduction in lead emissions was reported. However, nine urban areas have yet to meet the federal airborne lead standard.[22] In groundwater lead is picked up from soil and rock minerals that the water flows through. The concentration of lead in groundwater is generally below 10 ppb. Any level above this could be indicative of contamination.[23]

Health Effects of Lead in Drinking Water

Since man began mining lead over 2,000 years ago, it has been a health problem. Earlier civilizations like the Roman Empire experienced lead poisoning, presumably from their food and drink containers and lead plumbing systems.[24] It is speculated that health effects from lead ingestion may have been a factor in the breakdown of the empire.[25]

Lead workers during the industrial revolution had a high rate of reproductive disorders. Spontaneous abortion, stillbirth, and premature delivery were high in female workers and wives of male workers. There was also a high infant mortality rate among their children.[26]

Lead should be considered a top priority pollutant by everyone considering its widespread occurrence in plumbing systems. Studies have shown a direct association between lead levels in the blood and lead levels in drinking water.[27] Further, studies have shown that decreasing lead levels in drinking water is subsequently accompanied by a decrease in lead levels in the blood.[28] A study in Glasgow, Scotland found that those who lived in homes built before 1945 had a measurablely higher level of lead in their blood and their water than those in homes built after 1945.[29] This indicated that much of the lead came from domestic plumbing. After treatment of the water supply with lime to increase the alkalinity, the lead levels in the blood of the residents dropped considerably.

Water is one of the most bioavailable routes of lead exposure in man. It is believed that lead in tap water may be more available for absorption or uptake by the human digestive tract than the lead found in foods.[30] Lead bound within food is not as easily available for uptake into the body from the intestines.[31]

Exposure to lead in drinking water is a serious threat to public health. Although lead is classified as a probable carcinogen at higher levels, the greatest concern is for non-cancer health effects which occur at very low levels. Exposure to low lead levels has been associated with reproductive dysfunction, fetal damage, and delayed neurological and physical development in children; and high blood pressure, heart attacks, kidney damage, and strokes in adults.[32]

It has been shown that mentally retarded children have higher blood lead levels than normal children. A Scottish study measured water lead levels in the homes where 77 mentally retarded children spent their first year of life, and compared them to water lead levels in the homes of 77 non-retarded children. Water lead levels were found to be significantly greater in the retarded children's homes. In addition, the likelihood of mental retardation was largely increased when lead levels were over 800 ppb. Lead levels in the blood of the retarded children were also considerably higher.[33]

The current evidence indicates that lead absorption through the intestine of newborns is considerably greater than in adults. Also, children have developing central nervous systems that are particularly vulnerable to damage from lead exposure. Lead has the ability to cross the placenta to the fetus. Consequently, in addition to the vulnerability of infants and children, pregnant women should be concerned about lead levels in water as well.[34] This is recognized as the most significant toxic effect of environmental lead exposure. Researchers reported in the **New England Journal of Medicine** that even low levels of lead can be a problem for pregnant women who can transfer it to the fetus. The researchers studied the relationship between prenatal lead exposure (levels were measured in umbilical cords at birth) and early cognitive development. At various stages of growth and development, infants in the high prenatal exposure group scored lower than the infants in the low exposure group. The study concluded that lead levels far below those identified by the Centers for Disease Control (CDC) as acceptable can negatively affect the fetus.[35] Tests on

66 sudden death syndrome infants compared to 23 infants that died in accidents (all 4–26 weeks old) discovered 68.5% more lead in the ribs and 43.9% more lead in the livers of the sudden death syndrome infants than the accident victims.[36]

Testing for Lead

The type of pipe or solder may not be clearly apparent, though inspection of the pipes and solder may be of some help in revealing its type. Lead pipe is a dull gray color, and can be scratched with a file. Newer copper is about the same color as a new penny. Pre-1930 homes are most likely to contain lead pipes. Up through the early part of this century lead pipes were routinely installed in new homes. Questioning the water company will provide information on whether service or main lines are lead. Installation of service lines in some locations, until recently, were made using lead pipe.[37] Replacing all the lead pipe in the house may not make a whole lot of sense if the service and main lines are all lead.

It would be highly advisable to test any drinking water for lead, regardless of the type of plumbing material it passes through. What type of pipe or solder extends into the well? What about the service line from the house to the street where it connects to the distribution system? What about the distribution system itself? Although plastic pipe made in the United States does not contain lead, some foreign manufacturers may use lead as a stabilizer.[38] *Only testing will reveal the lead level in the water,* and consequently the extent of exposure to lead.

Test Results and Maximum Contaminant Level (MCL) for Lead

At this writing, the Maximum Contaminant Level (MCL) for lead in drinking water is 50 ppb. Under the Safe Drinking Water Act of 1986, water companies are required to notify customers of any excess lead levels in their tap water and provide information on its health effects. Municipal governments which control almost all the nation's community water systems would be responsible for complying with this reporting rule.

This current lead standard is generally considered too high to reduce the risk of health effects. Originally, it was proposed to reduce the acceptable lead level in water from 50 ppb to 20 ppb. Currently, it is proposed to lower the level even further, to 5 ppb. The cost of bringing drinking water supplies into compliance with the 20 ppb level was estimated at $3.80 per person per year.[39] According to the EPA, an estimated 40 million Americans may be using water that contains over 20 ppb of lead.[40]

Reducing Lead in Drinking Water

The contact time of the water with pipe or solder containing lead is probably the most important factor determining the amount of the lead that leaches into the water. The less contact with lead pipes or solder, the lower the likelihood of high lead levels in the water. The longer the contact with lead pipes or solder, the higher the level of lead in the water. The highest lead levels will be in the first water used in the morning to make coffee or prepare breakfast, etc. This water has remained in the pipes overnight in constant contact with surfaces of the piping. Water used during the day may also contain measurable levels of lead, but not at the increased levels found initially. So, as a risk-lowering measure, flush the faucets, especially in the morning and after stagnation times throughout the day. Also, don't forget to let it run a while when returning from vacation. This level would probably be extraordinary.

The procedure for flushing faucets is as follows. Allow the water to run for a short period after a change in temperature to colder water is felt. This colder water will represent water that stood in pipes outside the foundation. Letting the water run a few additional minutes after the colder temperature is felt should assure that both the house plumbing and the service line connected to the main or well have been flushed of stagnant water. Flush each faucet.

To conserve water, take a shower in the morning before flushing the faucet to be used for drinking water. This flushes the whole-house plumbing without wasting water. After the faucet is flushed, fill a glass container and put it in the refrigerator for drinking and cooking throughout the day. This eliminates the need for reflushing of the plumbing throughout the day when water is needed.

Flushing *will* reduce lead levels. By how much is variable. One *must* *test* the water to know how effective flushing is in reducing lead levels. In apartment buildings or high rises, testing is particularly vital because flushing the pipes is not likely to appreciably reduce levels due to the huge plumbing systems and large diameter pipes connected by lead solder.[41]

When moving into recently built homes, remove strainers from faucets and let the water flow for 15 minutes to remove loose lead solder or flux from the pipes. Then periodically check strainers for lead particle accumulation. Those in houses less than five years old are especially at risk of high lead levels in water. If the water is not seriously corrosive, deposits from minerals may eventually coat the inside walls, protecting the water from the solder. During the initial five years, however, the water generally comes in direct contact with the lead.

A number of factors contribute to lead concentrations in tap water, so a water treatment device may not be the "cure-all." Home treatment methods like calcite filters to reduce corrosivity (must be installed in the line *before* any lead connections) or lead removal by reverse osmosis (RO) devices are options to think about. When the initial and ongoing costs of treatment systems are considered, it may be wise to look at the replacement of lead pipes or solder as a permanent solution, particularly in the case of corrosive water. After comparing, it may be more economical or effective to simply to flush the lines thoroughly before use. Any decision will depend on the situation as well as a complete and accurate lead water analysis. Always test after the installation of treatment devices or filters to make sure they are doing their job.

It is unwise to soften water used for drinking. The use of water softeners in hard water areas can result in significant corrosion of certain types of metal piping material, like lead.[42] Needless to say, it is rather senseless for people to use softeners which make water more corrosive when municipalities are adding lime to increase hardness. When purchasing a home or putting in a new well, test to be sure the water is not excessively soft.

Consider bottled water for protection against lead if the water is corrosive and the home is new. This would also be a good idea if there is a young child or pregnant woman in the home. According to Michael Cooke, Director of the EPA Office of Drinking Water, even a few months of exposure to lead in the case of an infant or pregnant woman may be a detrimental factor in the development of the child.[43]

Studies of treatment-plant scale RO units have shown they are capable of removing 95% of the lead from water containing 100 ppb of lead.[44] One disadvantage is that RO softens the water by removing dissolved solids.

ARSENIC

Most people associate arsenic with murder mysteries, though many do not realize it is technically a metal. Arsenic is an acute poison, however, the concentration necessary to cause acute effects is extremely rare in the environment. The concern for arsenic in drinking water involves chronic or long-term exposure. The level which will prevent chronic health effects is the subject of debate.

Sources of Arsenic in Drinking Water

Arsenic is present in copper, lead, zinc, iron, manganese, uranium, and gold ores. It is released during industrial activities like smelting when, as an impurity, it is separated from the copper, lead, or zinc. Significant accumulations of arsenic (as well as many other metals) have been found in the soil in the vicinity of smelters. Arsenic emission to the atmosphere is a consideration at smelting facilities. In Nova Scotia, high levels of arsenic in surface and well water were associated with gold mining.[45] Waste rock and mine tailings were implicated as the cause of surface water contamination.

Other uses of arsenic include: pesticides on cotton and tobacco; wood preservatives; sheep dip; and a livestock feed growth-aid. High-purity arsenic is also used in semiconductors. Coal, particularly anthracite coal from the eastern U.S., contains a lot of arsenic which is released during combustion. Although still used in a wide variety of industrial processes, including glass making and feed additives, its uses for many processes like paint pigments and arsenical pesticides have diminished with the rise in SOC usage (Table 7-3).

In Minnesota in 1972, arsenic levels of 21 ppm and 11.8 ppm in drinking water were detected on separate occasions from a recently drilled well at a warehouse where workers were experiencing nausea, vomitting, abdominal pain, and diarrhea. Samples of the soil were taken near the warehouse at the surface and at a depth of two meters. Particles were found in the soil which contained 12,600 ppm of arsenic (yes, that is a lot). Through interviews with area residents it was discovered that in the 1930s there had been a problem with grasshoppers and arsenic had been used as a pesticide. It was mixed with bait which apparently had been stored on the open ground and retrieved by simply shoveling it as needed. It is surmised that the bait had been buried when it was no longer required.[46]

Although found in surface and groundwater, arsenic levels are typically higher in groundwater. Limited surveys have found 17% of groundwater supplies sampled, and 2% of surface waters sampled had arsenic levels over 5 ppb.[48]

Health Effects of Arsenic in Drinking Water

Arsenic can exist in two different forms in water, depending on how it bonds with other elements. Arsenic III, the trivalent form, is more toxic and tends to accumulate in the body.[49] Arsenic V, the pentavalent form, is less prone to reactions with gastrointestinal tract membranes, and is more easily excreted from the body.[50]

Carcinogenicity from chronic exposure to arsenic has contradicting evidence. Existing data however, points to arsenic as having mutagenic

Table 7-3
Sources of Arsenic[13]

- lead & zinc smelting operations
- nonferrous metal ore mining & milling
- phosphorous production
- water & wastewater treatment
- iron ore & steel making

- energy production†
- arsenic production
- copper refining
- copper smelting
- manganese production

† In energy production, emissions from coal are considerably greater than from petroleum or natural gas because arsenic concentrations are greatest in coal.

and teratogenic properties.[51] A study of chronic human exposure to elevated arsenic levels in drinking water and foodstuffs in Chile found an increase in skin lesions, clotting or restricting of major arteries, and myocardial infarction.[52] It was concluded that chronic arsenic poisoning could be a factor in myocardial infarction in persons under 40. Dr. Philip Landrigan, Director of Occupational Medicine at the Mount Sinai School of Medicine in New York, notes that two out of the three kinds of skin cancers caused by arsenic ingestion can be fatal.[53] Consumption of arsenic is also associated with other health effects like liver cancer, cardiovascular disease, and disorders of the central nervous system.

The United States Public Health Service (USPHS) in 1962 set a level of 0.05 ppm for arsenic in drinking water, advising that continual consumption of water above this level could be dangerous. It has been reported that 63 community water supplies have total arsenic above the 0.05 ppm level.[54] Certain areas of the U.S. (and the world) have natural levels of arsenic in well and spring water that exceed this level. The theory exists that while arsenic is an acute poison, it is also an essential element to humans at low levels. The city of Fallon, Nevada has been testing for arsenic since the 1960s and reports a consistent level of 0.10 ppm with little variation. It is believed that this was also the level in 1941 when they began using the aquifer. The city has reported no adverse health effects in the population as a result of the arsenic level, and even contends that arsenic may have an anticancer effect.[55] According to tests, the arsenic in Fallon's water is 100% pentavalent (Arsenic V).[56]

According to the National Academy of Sciences (NAS), the most important shortcoming in our knowledge of arsenic concerns its fate in the environment.[57] Since very little information is available on its fate in nature, it is not possible to accurately determine if a potentially dangerous accumulation of arsenic is occurring in any portion of the biosphere. This leaves us with quite a bit of uncertainty regarding a safe level for arsenic in drinking water.

Testing for Arsenic

No tests are commercially available to distinguish between the different forms of arsenic. Only a test to determine total arsenic is available. Sample as described in the general metals section. Request that arsenic analysis be included in the metals test.

Test Results and MCL for Arsenic

The EPA is currently proposing a level of 0.05 ppm for arsenic in drinking water due to the potential health effects and because it seems to be widespread. This is creating a lot of backlash and negative response from water treatment companies due to the difficulty of removing arsenic from water. According to some water treatment companies, the cost to remove even minute levels is excessive. Additionally, they claim the cost/benefit comparison of such a rule does not hold up when the levels found in some communities show no adverse health effects.

Reducing Arsenic in Drinking Water

Small distillation units are available to remove arsenic. Bottled water for cooking and drinking is also an alternative.

CADMIUM

Cadmium is in the same category as lead and arsenic in its potential as a health hazard. Cadmium, like other metals, exists naturally in minerals. Again, its extraction from ore for production and usage have made it bioavailable.

Sources of Cadmium in Drinking Water

The main industries using cadmium are: electroplating; pigments; PVC and plastics manufacturing; nickel-cadmium battery production; and the manufacture of alloys and solder (Table 7-4). Other avenues for its introduction into the environment are: mining and smelting operations; scrap metal reprocessing; incineration of plastics containing cadmium; fossil fuel emissions; application of fertilizers; and industrial sewage sludge disposal.[58] Pigments, plastics, and nickel-cadmium batteries contribute comparatively less cadmium to the environment.[59]

The leaching of cadmium from landfills is also considered to be a significant source with obvious environmental consequences. The great majority of the cadmium produced is deposited on land as waste. Water contamination by cadmium can result from decomposition of landfilled products containing cadmium, like batteries.

Municipal water treatment facililties which incorporate sand filtration into the treatment system remove over 80% of the cadmium present. Those facilities that do not utilize sand filtration cannot appreciably remove cadmium, leaving 80–90% of the cadmium in the water.[60] As in the case of lead, cadmium levels in tap water appear elevated in areas with soft, acidic water. Presumably, the cadmium comes from solder, plated plumbing fittings, or galvanized iron piping materials.

Table 7-4
Sources of Cadmium[61]

- mining & smelting of zinc, cadmium, lead, & copper ores
- pigments in paints, plastics, enamels, lacquers, & printing inks
- nickel-cadmium batteries
- sewage effluent & sludge disposal
- plastic stabilizers
- alloys
- phosphate fertilizers
- urban runoff
- electroplating processes
- atmospheric deposition

Health Effects of Cadmium in Drinking Water

Not surprisingly, exposure to cadmium has increased in this century.[62] The most publicized occurrence of cadmium poisoning happened along the Jintsu River in Japan where people drank water (and ate rice grown in the water) contaminated with cadmium from mining operations. Due to cadmium's effect on red blood cells, affected patients experienced bone fractures and severe pain even after only slight movement. Hence, this disease became known as "itai, itai" (translated as "ouch, ouch") disease.

Cadmium is also associated with hypertension; effects on the lungs, cardiovascular system, and central nervous system; liver and kidney damage; and mutagenic, teratogenic, and fetotoxic effects. The kidney appears to be the main target organ of cadmium in the body. The most notable effect of chronic cadmium exposure, as reported in scientific literature, is irreversible renal tubular dysfunction.[63] The International Agency for Research on Cancer (IARC) lists cadmium as having limited evidence of being a carcinogen to humans.[64]

A study at Texas A & M University headed by Dr. Jack Nation reported that cadmium caused anxiety in rats. The rats, which typically reject alcohol, learned that alcohol reduces anxiety. The study rather surprisingly suggested that there may be a link between pollution problems in the United States and alcohol abuse.[65]

Testing for Cadmium

Follow the general metals sampling instructions. Request that cadmium analysis be included in the metals test.

Test Results and MCL for Cadmium

The EPA has proposed a maximum level of .005 ppm for cadmium in drinking water.

Reducing Cadmium in Drinking Water

Flush water lines before using water for cooking or drinking as described in the lead section. Municipalities should treat water to make it harder and less acidic. Home treatment may include RO or distillation.

COPPER

Copper is generally of lesser concern than the metals discussed previously. It is considered an essential element. At low levels it is required for many enzyme reactions within the body.[66]

Sources of Copper in Drinking Water

Sources of copper are: smelting and refining industries; copper wire manufacturing; coal burning; and iron and steel industries. Copper is found in industrial discharge and is used as an algaecide in reservoirs.[67] As seen in Table 7-5, most uses of copper are related to electrical and plumbing fixtures. Plumbing fixtures, specifically copper and brass, are the main source of copper in drinking water. Hardness, acidity, and alkalinity are again factors in the degree of leaching of the metal from the pipes. In areas with soft, acidic water, copper concentrations may be increased in drinking water from copper pipes in much the same way as lead.

Table 7-5
Uses of Copper[68]

• communication, magnet, & building wire	• roofing
• electrical switches & printed circuits	• heat exchangers
• plumbing and fittings	• foil
• valves	• fungicides
	• busbars

Health Effects of Copper in Drinking Water

A family in Vermont was documented as experiencing vomitting and abdominal pain within minutes after drinking water, or juice and coffee made from water, containing elevated levels of copper.[69] They reported the water as having a blue tint. A median level of 3.07 ppm was detected in the water although an early morning level of 7.8 ppm was determined. These levels were considerably higher than other homes in the community. It was concluded that the reason for the high concentration of copper was the aggressive character of the water in the area (soft, low pH), and the fact that the house was located at the end of a copper water main (the last house on the street). All the copper which dissolved along the main ended up their water. The main was replaced by an iron, cement-lined pipe and copper levels dropped to less than 1.0 ppm. The chronic effects of copper poisoning are not known.

Testing for Copper

Sample as per instructions for metals testing. Request that copper analysis be included in the metals test.

Test Results and MCL for Copper

Copper is not currently regulated in drinking water, however, the EPA is proposing a level of 1.3 ppm due to consideration of its gastrointestinal disturbances, its possible link to liver and kidney damage, and its anemic effects in cases of acute poisoning.[70] Copper is regulated in ambient (non-drinking) water at 1.0 ppm due to its ability to stain laundry and plumbing fixtures.

CHROMIUM

Chromium exists generally in two forms in water: chromium III (CrIII) and chromium VI (CrVI). Large amounts of CrVI are produced and used by industry in this country. Due to its high water solubility, compounds containing CrVI are often found in natural waterways.[71] CrVI is also more toxic to humans than CrIII.

Chromium readily converts back and forth between the CrIII and CrVI forms. Whether chromium is in the CrIII or CrVI state will depend on surrounding conditions like pH, the amount of dissolved oxygen, and other factors.[72] Some areas in the country have reported elevated levels of CrVI in groundwater. These areas are generally alkaline (high pH) and have conditions which promote oxidation.[73] Under certain conditions, all the chromium will be in the CrIII form but, if the conditions change, all the chromium can be converted to the CrVI form or a percentage of each, say 60% CrVI and 40% CrIII.

Unfortunately, CrVI is the most likely form in which chromium exists in natural waters. In addition, during water treatment (mainly chlorination), oxidation will transform any CrIII to CrVI. Testing at the plant to determine chromium levels in municipal systems does not account for any further oxidation between the plant and the tap. The residual treatment capacity within the system will keep chromium in the more dangerous CrVI state. So if chromium is present initially, then by the time it reaches the faucet the chromium will likely be predominantly CrVI.[74]

Sources of Chromium in Drinking Water

Sources of chromium include old mining operations and plating wastes. Industries most likely to generate chromium wastes are involved in the manufacture or use of: chrome alloys; chrome plating; corrosion inhibitors; oxidizing agents; chromium compound manufacturing; pigments; photographics; textiles; ceramics; glass; anodizing aluminum; linoleum; paint (yellow); explosives; batteries; rubber tires; and paper.

Landfill disposal of CrVI waste is considered a serious potential hazard because of chromium's great mobility in soil material. Industrial wastes are often very acidic, particularly those from metal plating operations where one might expect chromium to be found. If put in landfills it may change the chemical makeup of leaching water. Landfills and disposal impoundments also provide ideal circumstances for CrIII to change to CrVI — alkaline conditions (high pH) and an oxidizing atmosphere.

Health Effects of Chromium in Drinking Water

Chromium III compounds are considered relatively harmless, and are believed to be an essential element for humans at low doses. CrVI, on the other hand, is considerably more toxic and has been linked to liver and kidney damage, internal hemorrhaging, respiratory disorders, and mutagenic effects.[75] CrVI has also been implicated in digestive tract cancers, although conclusive information is lacking.[76] The IARC has determined that there is sufficient evidence that CrVI causes cancer in humans and animals.[77] Its toxicity is believed to be enhanced by the fact that the CrVI is more water soluble and can cross body membranes more easily. It is therefore taken up by body tissues more readily. Researchers find it hard to pinpoint the level at which chromium exhibits toxic effects. Observed levels of toxic effects are difficult to precisely measure.

Testing for Chromium

Analytical methods used by most laboratories do not specify or differentiate between CrVI and CrIII. A method does exist to calculate only the more harmful CrVI, however, there is some debate about its sensitivity and not all labs may perform this test. In addition, if one uses municipal water it is quite probable that if there is chromium in the water, it will be in the CrVI form anyway. So the conventional test for total chromium (CrIII and CrVI) will be just as suitable. For most purposes this makes things much neater, is less confusing, and would generally be less expensive.

Sample as described for all metals testing. Request that chromium analysis be included in the metals test.

Test Results and MCL for Chromium

The EPA has proposed a level of 0.12 ppm for total chromium in drinking water based on the toxic effects of CrVI.

MERCURY

Mercury exists in primarily two forms: elemental (inorganic) and methyl (organic). It is most dangerous in the organic form which allows it to cross body membranes readily and become absorbed by the body tissues. The mercury discharged from industrial plants is in the inorganic form, but bacteria in sediments at the bottom of waterways can transform it into the more dangerous methyl-mercury.

Sources of Mercury in Drinking Water

Certain industrial activities that one would not usually associate with mercury, fossil fuel burning plants and metal smelters, actually discharge significant amounts of the metal into the environment. Other sources of mercury are: batteries; electrical switching equipment; and measuring and monitoring instruments, i.e., medical and lab equipment, and thermometers (Table 7-6).

One of the most famous cases of mercury poisoning occurred in Minamata Bay, Japan. Chemical plants using mercury as a process catalyst and dumping it into the bay were responsible for 43 deaths and a number of birth defects after people ate contaminated fish.[79] Mercury is one of the metals that bioaccumulates. This means that the concentration of mercury in the fish population will be many times higher than in the contaminated water. For example, suppose a water source has a concentration of 1 ppm due to contamination. Fish in the water could contain 10 to 100 times that amount. The amount of accumulation depends on, among other things, the species of fish. Tuna and swordfish are the prominent accumulators, but most animals (including humans) will also accumulate mercury. What does all this have to do with drinking water? Food is

---

Table 7-6
Uses for Mercury[78]

- electrical conductor used in the manufacture of light switches and wires, mercury vapor lights, and certain batteries
- intermediary in plastics and organic chemical manufacture
- paint preservative and mildew inhibitors
- preservatives in paint, fabric, cosmetics, and pharmaceuticals
- wood preservatives (found in pulp and paper wastes)

---

considered the main source of exposure to mercury, and the primary dietary source is apparently fish. So, in typical instances, the amount of mercury in drinking water is negligible in comparison to the average diet.

Elemental mercury is used in some equipment, namely mercury-sealed well pumps, in drinking water supply wells. There has been some concern regarding the possibility of accidental contamination.[80] Some scientists argue, however, that conditions are not conducive to the transformation to methyl-mercury (not enough oxygen in wells) and the mercury would not pose a health risk because it is not concentrated as in the case with contaminated fish.[81]

Health Effects of Mercury in Drinking Water

All mercury compounds are considered poisonous, especially methyl-mercury which is very fat-soluble and easily absorbed by the body. The elemental mercury metal is not as toxic because it is generally insoluble and not absorbed very well by the body. The vapor, however, is very dangerous.

Methyl-mercury poisoning causes central nervous system damage. However, inorganic elemental mercury usually ends up in the kidney. Surprisingly, miners at an old mercury mine in Almaden, Spain, reportedly drank elemental mercury for centuries to relieve constipation.[82] This, of course, is not advisable, but it shows that methyl-mercury is the worst of the two.

Testing for Mercury

Ask the laboratory for specific sampling instructions and bottles. Mercury requires a special testing procedure different from the other metals, called Cold Vapor. It utilizes the same instrument as the other metals, but the method is different and therefore more expensive. It also does not differentiate between elemental and methyl-mercury.

8. BIOLOGICAL CONTAMINATION: Waterborne Diseases

*"And then they dipped and drank their fill
of water fresh from mead and hill."*
— Sam Walter Foss (1858–1911)

Hunter/gatherer man of millenia past had little negative effect on water or other aspects of the natural world. Human and animal populations were more proportional to one another, the food chain was well balanced, and the water necessary for life's existence was clean and seemingly infinite. The waste of animals and man was a nutrient source in the natural cycle. The microorganisms and parasites in the waste were a necessary and even a vital part of the biological cycle. Even as primitive man settled into early farming practices and became more stationary, the biological balance remained, and contamination was not a problem. The effect of a few people on the ecosystem was not nearly as detrimental as the effect of a village or town.

Since water is a requirement for life, groups of people settled near waterways and ended up polluting the very source on which they depended. As people gathered in cities, the problem of human waste disposal became a health probelm. Cholera, typhoid, and other waterborne diseases began to take their toll on humans as nature attempted to control the population.

The larger the population, the greater the chance of being infected by pathogenic parasites that are cultured or proliferated within the human body. The human-parasite cycle had begun.

An association between certain infectious diseases and drinking water contaminated with sewage was finally made by Dr. John Snow in 1854 in his famous study of London's Broad street pump.[1] He noted that peopled afflicted with cholera were clustered in one area and isolated the Broad Street pump as the source used by those affected. As late as the U.S. Civil War, the concept of waterborne disease was still poorly understood. Encamped soldiers often disposed of their waste upriver but drew drinking water from downriver, resulting in widespread dysentery.

Today water treatment methods are used to prevent the spread of disease that once helped regulate our population. In 1900, thirty people out of every 100,000 in the United States died of typhoid. By 1907 water filtration was practiced by most cities, and by 1914 so was chlorination.[2] This led to a drop in the number of outbreaks of waterborne diseases in the latter part of this century. In Cincinnati the yearly typhoid rate of 379 per 100,000 people in the years 1905–1907 decreased to 60 per 100,000 people between 1908 and 1910 with the beginning of sedimentation and filtration treatment. The introduction of chlorination after 1910 decreased this rate even further. The national typhoid death rate in the United States between 1900 and 1928 dropped from 36 to 5 cases per 100,000 people.[3]

What Causes Waterborne Disease?

A definate difference exists between bacterial and chemical pollution. Much of the wastewater from industry is highly polluted chemically, but contains no biological contaminants. In the same way, domestic sewage might pass chemical tests but still be contaminated with microorganisms.

Modern sewage treatment methods merely reduce bacterial numbers, they do not remove all bacteria. Usually the discharge of treated sewage to rivers, which are themselves self-cleaning, can result in the reduction of microoganisms to undetectable levels. This is provided the level is low enough initially and the receiving waterway is not already a biologically-stressed system as a result of other discharges. If the discharge of sewage is constant, increased levels of bacteria may override a river's cleansing capability. Adequate dillution of sewage wastes into the receiving natural waterway is necessary to ensure that natural processes of self-purification can occur. Often, however, after heavy rains wastewater treatment systems are not able to handle the huge volume of water. Consequently, raw sewage is discharged directly into waterways.

Waterborne diseases are caused by very small (unseen by the naked eye) organisms originating from animal or human feces. Although they do wreak havoc with the human gastrointestinal system, waterborne diseases are rarely fatal today as they were at the turn of the century when cholera and typhoid were prevalent. In current outbreaks of waterborne disease, effects are usually acute, not chronic as in the case of most chemical contamination. An incubation period, the amount of time after the water is ingested before symptoms begin to show, is typical. During this time the bacteria or contaminant organisms multiply to a level which will cause illness. The incubation period can be anywhere from a few days to a few weeks.

Whether or not an infection occurs will depend on factors like how much and what type of bacteria or virus is present. For certain types of bacteria only a small number of cells are required to initiate infection. With other types it is the opposite. Bacterial contamination does not automatically mean that illness will result. Nonetheless, it is reasonable to assume that the presence of pathogenic or disease-causing microorganisms and the accompanying health risk increases along with the bacterial count. The presence of any pathogen is a clear indication of a health risk.[4] However, the presence of a pathogen does not necessarily mean disease is eminent. All those exposed may not suffer from illness. There is not always a clearly identifiable cause and effect relationship between microbiological organisms and illness.[5]

Cases of waterborne disease averaged about 9,000 per year between 1975 and 1984.[6] However, it is believed that the great majority of waterborne disease cases go unreported.[7] According to the EPA, the actual number of cases of waterborne disease may be ten to several hundred times higher than those reported.[8] Most states do not have an adequate surveillance and reporting system. An outbreak in an area may not be recognized as such, and drinking water as the source of illness in a community may be overlooked. Consequently, the reported number of outbreaks may only be a fraction of the total.

Those suffering from the effects of waterborne disease may not even realize drinking water is the cause.[9] A specific cause was determined in only 48% of the reported waterborne disease outbreaks between 1971 and 1982.[10] When the specific agent cannot be identified, the official report will list the cause as gastroenteritis or "acute gatrointestinal illness of unknown origin."

The reason for the difficulty in determining the cause of waterborne disease outbreaks is the incubation period. By the time those infected are experiencing symptoms, the quality of the water source may have changed and the infectious microoganisms may not remain. By the time people get sick the water that caused the illness is at the sewage plant or in the river. Many times in larger outbreaks, human samples (blood, urine, stools) are taken in order to determine the infecting contaminant because the common water source of those infected is determined to be safe after testing.

Another difficulty in identifying and reporting waterborne disease is that many outbreaks occur in transient populations, which use campgrounds or restaurants. Vacationers and travelers will be miles away by the time symptoms occur, making it difficult to trace the source of the infection.

Sources of Contamination in Municipal Water Systems

Biologically safe water is a top priority among government regulators and most municipal systems take every precaution to provide it. After all, biological treatment was the specific function of treatment plant design. Adequate overall treatment, including disinfection, is the only protection between the raw sewage discharged upstream and the drinking water taps downstream.

Community water system outbreaks generally occur in small water systems and result in a limited number of illnesses. In these systems, untreated or inadequately treated water is responsible for the majority of waterborne disease outbreaks. Too many people in one area will sooner or later be confronted with disease resulting from too much waste. With adequate treatment of waste and drinking water, this can be overcome. Any breakdown in the treatment system puts us back at the mercy of natural laws.

Residual chlorine, left over after disinfection, must be maintained in the *distribution system* (Figure 8-1) to eliminate the growth of microorganisms. Water initially exiting the drinking water plant may be disinfected, but growth that can lead to outbreaks is still possible in the system due to:[11]

- contamination of the storage unit for the distribution system (the water tower, for example);
- stagnant water in dead-end pipes where microorganisms can grow;
- lack of adequate residual chlorine sustained throughout the system;
- an abundance of nutrients for the growth of microorganisms;
- seasonal temperature changes;

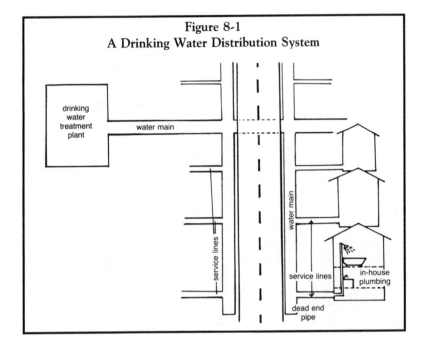

Figure 8-1
A Drinking Water Distribution System

- sediment deposits which allow for stagnation; and
- stagnant or standing water.

Sudden contamination, which may be more isolated than treatment deficiencies, may result from plumbing cross connections, reduction in line pressure (backsiphonage), line breaks, and repairs to the distribution system. [12]

Public water systems may have dead-end lines where low flow can be a factor in the growth of bacteria which has survived chlorination. Flushing is the prescribed remedy and many utilities periodically flush their lines. On a smaller scale, faucet aerators are also good environments for the buildup of organic deposits that foster bacteria growth. Always allow the water to run for a minute or so after periods of non-use. High bacteria levels should be expected in water that has been stagnant or has remained in pipes overnight. The first water out of the pipes in the morning will likely have the highest bacteria levels.

Sources of Contamination in Groundwater Systems

Individual wells and small drinking water systems utilizing groundwater are seen as being at the greatest risk for contamination. [13] These systems generally use groundwater without treatment, and this lack of treatment is considered one of the major factors in waterborne disease outbreaks and illness. Between 1971 and 1982, 51% of the reported outbreaks were due to contaminated groundwater that was either inadequately disinfected or not disinfected at all.[14]

Microbiological contamination of groundwater occurs most often from septic tanks or cesspool overflow or drainage. Placement of septic tanks in soils unsuitable for drainage, or in areas with a high water table that can flood out the drain field, is only inviting problems. Groundwater pollution often occurs when septic tanks are used in areas where they overlay a water table that rises and falls seasonally.[15] A minimum two meter depth is recommended between the septic tank absorption field and the water table.

Incorrect drinking water well construction has also been blamed for contamination. Improperly cased wells can be affected by contaminated groundwater from other aquifers. Poor casings can also allow contaminated surface water runoff to enter the well.

Other unforseen potential sources of contamination should also be considered. Dredging a river can expose a well field and contaminated water from the river can enter drinking water wells. Domestic animal feedlots, where a dense population of animals exists, are also a major source of microbiological contamination of drinking water wells.

Certain factors aid in the movement of the water contaminated with human or animal waste. Fractured or channeled bedrock can contribute to the random and unpredictable flow of biologically polluted waters from the surface into groundwater. Overpumping of wells by several homes in an area can increase flow and lateral movement of groundwater enough to move bacteria long distances. What had once been considered safe distances between wells and waste effluent systems may not be in all cases.[16]

Prevention is the best defense against biological contamination. Alternate water sources or treatment methods will not be necessary if the common sources of human and domestic animal wastes can be avoided. Unlike the often hidden sources of chemical pollutants, sources of biological contaminants are usually easier to identify.

TESTING FOR MICROBIOLOGICAL CONTAMINANTS

It is unlikely that water testing will prevent an isolated case of waterborne disease in a municipal water system. The EPA is proposing strict regulations for many municipal water companies regarding testing for pathogens, treatment to eliminate waterborne pathogens, and public notice of violations. For most municipal systems, particularly those using surface water sources, the microbiological quality is constantly changing as large amounts of water are transported and consumed. But, small groundwater systems like private wells change much less rapidly in their overall quality. For these, there are simple, low cost tests which will reveal the general microbiological quality of the water for an extended period of time.

Total Coliform

Many of the infectious microorganisms found in water are difficult to isolate and identify. Therefore, their complete analysis would not be practical in terms of the time and money required. Consequently, the analysis of **indicator organisms** is commonly used. The indicator organisms used in determinining the presence of pathogenic bacteria in water are called **coliforms.** Coliforms are found in the excrement of man and warm-blooded animals and accompany infectious or pathogenic organims. If coliforms are found in the water, it is very likely that pathogenic bacteria will also be present. The total coliform test can be done much faster, more easily, and at much less expense than the battery of cumbersome tests that would be necessary to detect the many different types of pathogenic bacteria. Community water companies and government regulatory agencies all agree it is better to analyze water frequently with this simple method than to use a series of complex methods to monitor water on an irregular basis.

The total coliform test, although it has some shortcomings, has been used for most of this century as an indicator of the total quality of a water supply.[17] It is used by municipcal water authorities to ensure proper treatment, and in some areas is required when a home with a well is sold. A negative coliform test is considered indicative of safe drinking water, void of waterborne disease or pathogens.[18]

It is generally agreed that even though the specific health effects of microbiological contamination are poorly understood, the presence of coliform is cause for immediate concern and the source should be found. The ill, the elderly, the very young, and those otherwise in a depressed state of health are more susceptible to the health effects of microbiological contaminants. Particular attention must be paid to them if their water is found to contain coliform contamination. In any case, the occurrence of coliforms in drinking water should not be ignored.

The rural homeowner with well water could make better use of the coliform test than a customer on a community water system. Community water systems routinely analyze for total coliform and an individual's analysis should not be markedly different. For the rural homeowner, a periodic analysis should suffice, provided the water supply is not subject to pollution. However, if any changes are made with regard to human or animal wastes disposal, more frequent analysis might be considered. Also, severe flooding almost always causes an increase in the total coliform bacteria.

Two different methods are currently used for determining total coliforms in drinking water, the Most Probable Number (MPN) and the Membrane Filter (MF) techniques. Chances are most labs will use the MF procedure. It is the least expensive and requires less equipment than the MPN method. In addition, the MF takes 22 hours while the MPN takes 48. It is also the preferred method because of its precision, reproducibility, and speed. Its sensitivity can be reduced, however, by interfering turbidity and large amounts of non-coliform bacteria.[19]

Two presence-absence (P-A) tests are being considered as alternative methods. P-A testing can eliminate some of the traditional problems with coliform testing. The advantages of a P-A concept for total coliform measurement, regardless of the test method, include:[20]

1. an easier determination of coliform presence rather than coliform density;
2. less dependence on quick transit time; and
3. the elimination of calculation problems in determining the number of coliforms.

All biological methods are tricky just by the nature of microorganisms in general. False-negative results are possible because coliforms may be suppressed or overshadowed by high populations of other bacterial organisms. Additionally, analyst judgment and interpretation are much more a part of the total coliform test than with other analytical methods. Few routine bacteriological tests are actually performed by registered microbiologists.[21] Generally, lab technicians untrained in microbiology handle this test. This does not mean they are ill trained, it simply means that an experienced microbiologist may interpret the results differently.

Often, when testing is done by municipal water systems, a positive result is assumed to be a laboratory or sampling error and the water is then resampled until a negative result is obtained. No action is taken to find out how the contamination occurred or whether a problem exists in treatment or distribution.[22] Only persistent positive test results will result in corrective action. In fact, under federal regulations, municipal water companies are entitled to some positive results as long as they do not have too many in a certain time period.

A waterborne disease outbreak in a municipal system may not be prevented even if the total coliform test provides an alert. By the time the sample is taken to the lab, analyzed, and the results are returned, the water has already been distributed, consumed, and is likely at the waste-water treatment plant. It is too late for any type of corrective action at the drinking water plant. All microbiological methods are viewed to be inadequate in that they only give information on the quality of water of a few days ago, not presently.[23]

Total coliform monitoring in municipal systems is basically designed to identify treatment and distribution deficiencies that could cause outbreaks of disease. Pathogens can exist where coliforms are not detected, and outbreaks have occurred in water systems where standard coliform levels have not been exceeded.[24] Disinfection, along with other treatment methods, are necessary for total coliforms to be a good indicator of overall water treatment efficiency. For this reason, coliform monitoring in small systems where disinfection is the only treatment is considered ineffective.[25]

Compliance with federal regulations for municipal water companies is based on the presence or absence of total coliforms as opposed to a specific level or density of coliforms. Even though the presence of coliforms is a clear indication of a health risk, those on private water systems are not subject to these regulations. An estimate of the density of coliforms can be requested to determine the degree of contamination. Total coliform results (MF) are reported as **total coliforms per 100 ml**, regardless of the volume of water in the sample. Results may also be reported as:

1. *Confluent growth* (with or without coliforms). This means the entire membrane was covered with a variety of bacteria.

2. *TNTC (too numerous to count).* Too many colonies (greater than 200) means an accurate count cannot be made. This is reported as *TNTC per 100 ml* (with or without coliforms). In either case, resubmit a new sample.

One should be aware that coliforms are not adequate for indicating the presence of pathogens not related to fecal contamination. These include legionella, giardia, and viruses. These infectious microorganisms will be discussed in subsequent sections in more detail.

Fecal Coliforms

Although many types of coliforms exist naturally in waterways, the presence of *fecal coliforms* in drinking water is a strong indication of recent sewage contamination. According to the EPA, "The presence of fecal coliforms indicates that an urgent public health problem probably exists, since human pathogens often co-exist with fecal coliforms."[26] Testing for fecal coliforms is required for municipal water treatment companies if a sample is positive for total coliforms. For private well owners a similar course of action is, at the very least, highly advisable. From this a clearer picture of the extent of contamination and risks can be obtained. With the relatively low cost of both total coliform and fecal coliform tests, those with a private well may consider testing for both initially.

Heterotrophic Plate Count (HPC)

Heterotrophic bacteria are those that utilize and need organic material in order to grow and reproduce. This very broad category is comprised of the bacterial pathogens discussed in the total coliform section, as well as other non-harmful bacteria.

The laboratory test for this large class of bacteria is called the Heterotrophic Plate Count (HPC) or the Standard Plate Count (SPC). According to *Standard Methods for the Examination of Water and Wastewater*, considered by many to be the bible of water testing, HPC is the best method available for determining the general bacterial quality of a water source.[27] The EPA is considering requiring public water companies to monitor for HPC along with total coliform.[28]

HPC is a simple, often overlooked test used to measure the density of a substantial portion of the total bacteria population in the sample. This result will, of course, include the harmless bacteria. However, it is a direct measure of the bacteria level and not an indictor test as is the case with total coliform. A high result reported for this test may be a signal of poor water quality, particularly if total coliform supports the results. It is possible that an HPC test result could be high simply due to harmless bacteria. Therefore, it is a good idea to test for both total coliform and HPC bacteria. If both are high it is fairly certain there is a problem.

The HPC procedure has been used to determine bacterial colonization of carbon filters used in home water treatment. The growth of pathogens in Point-of-Use (POU) treatment devices is better determined by HPC than by total coliform.[29] This is because total coliforms are rarely found in the effluents of home treatment devices.

Another good reason to consider HPC analysis is that a high HPC bacterial density could overshadow or interfere with the total coliform method. This could give a false low result for total coliform. Therefore, a high HPC test result may indicate that the total coliform results should not be relied upon. The EPA is proposing a HPC limit of 500 bacteria per ml of water for municipal water systems that are required to test for HPC. This level can be used as a guide for those testing their well water.

SAMPLING FOR TOTAL COLIFORM, FECAL COLIFORM, AND HPC

Water samples taken for microbiological analysis should be tested as quickly as possible. Bacteria are unpredictable due to the many different factors like time, nutrient supply, and temperature, that can affect their survival and growth. Significant changes in bacterial densities over time are possible. Samples should be iced and analyzed within 6 hours to prevent changes in total coliform counts or bacterial densities.[30]

Nutrients present in the water could cause a large increase in bacteria levels during sample transport to the lab, especially at temperatures above 55°F. Alternately, if the sample is stored too long, the bacteria could exhaust its entire food supply and quickly die off in large numbers.[31] Consequently, the bacterial density could be significantly reduced by testing time. In either case, the result would be false or misleading.

Samples analyzed directly after they are collected will give the most accurate results. The maximum time allowed for sample transport and storage time is a debatable point. Generally, the sooner the better. Bacterial populations change more rapidly than coliform density, so HPC samples require more urgency. Research indicates that the results of HPC samples analyzed immediately and those tested eight hours later were very similar, regardless of whether they were refrigerated or not.[32] However, eight hours is considered the maximum time between sample collection and testing of samples that are not iced.[33] Currently, it is acceptable to hold samples on ice for up to 30 hours to allow for transport to the lab,[34] but it is recommended that drinking water samples be iced and analyzed promptly the same day.

Though not as critical, time is an important factor in coliform testing as well. Studies have shown sizeable changes in coliform numbers in some samples held more than 24 hours.[35] If the sample cannot be delivered to the lab immediately, put it on ice and try to get it there as quickly as possible. Prompt delivery is essential, so a nearby lab is best.

Obtain a sterile sample bottle from the lab. Make sure a preservative, EDTA and tetrasodium salt, is included to keep heavy metals from interfering with the test. If the water is chlorinated, a dechlorinating agent should be added to the sample bottle by the lab. Do not sample on Thursday or Friday, or before a holiday. The sample may arrive too late for someone to "prep" it for analysis and the required incubation time may not be met over a weekend or holiday. Do not put the samples in a hot trunk or the sun as this will have a definite effect on the bacterial count. Make sure the sample bottle will not become submerged in water when the ice in the transport container melts.

How to sample:
1. Remove the aerator or faucet attachment. Avoid taps with anti-splash units and home treatment devices, or remove them before taking the sample. These are ideal breeding grounds for bacteria, especially if they are not routinely cleaned. Storage or holding tanks can also harbor bacteria and contribute to their growth.

2. Allow the water to run for several minutes. The time varies with each specific situation, but the idea is to obtain a sample representative of the water in the main lines or the well. This will assure a representative sample rather than water standing in the pipes. Take great care to avoid touching the bottle neck. Also, do not hit it against the faucet or sink. This could cause contamintion of the sample and give a false result. Avoid taps that do not have a steady flow. A random or sporadic flow rate can dislodge bacteria from pipe walls erratically, giving a false impression of water levels of bacteria. Do not set the bottle cap down or touch the inside. Do not rinse or touch the inside of the bottle. It has been sterilized at the lab and would be contaminated by such actions.

3. Fill the bottle slowly, without splashing, to within one inch of the top. Put the cap on immediately, then turn off the water.

4. Mark the date and time the sample was taken on the side of the bottle.

TURBIDITY
Turbidity, the cloudy appearance of water, is caused by suspended material. It occurs most commonly in individual water supplies that use surface water sources. This suspended material may consist of algae growth, clay, microorganisms, natural silt, natural organic chemicals, decaying vegitation, and organic and inorganic industrial chemical wastes discharged to surface water.[36]

The reason turbidity is grouped with microbiological tests lies in its relationship to the pollution and treatment of water. For example, turbidity can:

• disrupt the disinfection efficiency in treatment processes;

- consist of toxic particles or particles that absorb or bond with toxic substances;
- interfere with the total coliform membrane filter (MF) procedure, giving incorrect results; and
- be an indication of the efficiency of the drinking water treatment process.

Public water supplies are required to treat their water to remove turbidity, and adequately treated surface water usually does not present a turbidity problem. Currently, the Maximum Contaminant Level (MCL) for turbidity is 1.0 NTU. Treatment for turbidity usually results in the removal of pathogens and some chemicals, especially those that adsorb onto particulates.[37] Since these particles may be harmful or may bond with harmful compounds, the removal of turbidity removes the threat or possibility of contamination. This reduces the health risks without having to distinguish between harmful and nonharmful particles. In areas where surface water is unfiltered, highly turbid water could be reaching the tap.

Little evidence exists to determine the direct impact of high turbidity on health. It is known that particles in water can carry hazardous chemicals and contribute to the formation of dangerous chlorinated compounds after disinfection. Also, high turbidity in contaminated water is usually accompanied by increased bacteria levels.[38]

The turbidiy test is an indicator test that will not give results on a specific pollutant. It will, however, provide information on the degree of overall contamination based on the level of particles. Turbidity measurement as well as turbidity removal in the treatment process takes into account all particulates, regardless of toxic effects.

Turbidity is not a useful indicator of the quality of groundwater because the majority of particles responsible for turbidity are removed by natural filtration as the water moves through the aquifer. Groundwater, however, can have a chemical basis for turbidity. Certain bacteria produce iron and sulfur solids and consequently, turbidity.

The measurement of turbidity is based on the fact that when light is passed through turbid water it is scattered, and the degree or amount of scattering is a measure of the amount of turbidity. The only method currently approved by the EPA is the Nephelometric Method. The results of the Nephelometric method will be expressed in Nephelometric Turbidity Units (NTU). If the decision is made to test for turbidity, make sure to request this method. Some labs are still utilizing the Jackson turbidity method. The Nephelometric measurement is well defined to limit variations in reading and results by different turbidimeters. The Jackson tur-bidity method is variable and has very limited applicability to drinking water measurements. Be wary of a lab running the Jackson turbidity test and converting the resulting JTU (Jackson Turbidity Units) to NTU by mathematical calculations. Having your results in NTU doesn't mean a thing when the analysis is incorrect or invalid.

GIARDIA

Giardia was inititally identified in 1681[39], but the first reported case of giardiasis in the U.S. wasn't until 1965 in Aspen, Colorado.[40] Giardia Lamblia is the specific microorganism responsible for giardiasis. It exists in the environment, usually water, in the cyst stage. When ingested, it can move into its reproductive stage and become infectious.[41] The cyst can survive in cold water for months. A recent study concluded that, contrary to previous theories, giardia cysts may be constantly present at low concentrations, even in isolated and pristine watersheds.[42]

Almost half of all waterborne outbreaks in community water systems are due to giardia lamblia, making it the most commonly identified patho-gen.[43] It is believed that the number of reported cases represents only a fraction of the actual number that occur. In addition, giardiasis outbreaks are increasing rapidly, doubling every five years.[44] Infections in the Los Angeles area are on the rise and no longer thought to be rare.[45]

Infection by giardia does not automatically cause illness. Many people and animals show no symptoms during a giardia infection. According to the Centers for Disease Control (CDC), finding giardia cysts in patient stools is not conclusive evidence that it is the cause of the illness.[46] It is not understood why some people get sick and others, though infected, shown no signs of it. The incubation period ranges from one to three weeks. Acute symptoms include gas, flatulence, explosive watery foul diarrhea, vomitting, and weight loss. In most people these symptoms last one to four weeks, but can last as little as three or four days or as long as a several months.[47]

Outbreaks have occurred throughout the U.S. but primarily in the New England, Rocky Mountain, and Pacific Northwestern states. In these mountainous areas where high quality water is expected, smaller, less comprehensive treatment plants are often found.[48] Most of these areas are mistakenly thought to be void of sewage or bacterial contamination.[49] Also, smaller systems do not always have the money that major systems have for treating water.

The majority of outbreaks are the result of consumption of untreated surface water, or water treated solely with chlorine.[50] Chlorination is sometimes the sole treatment method and only done to meet total coliform requirements. A giardiasis outbreak in Rome, New York in 1974, which resulted in 4,800 cases of illness, was blamed on lack of adequate chlorination.[51] Unfiltered surface water caused an outbreak in August 1983 in Red Lodge, Montana, where about 780 people were infected.[52]

Although giardiasis outbreaks occur most often in surface water systems, giardia cysts have been found in groundwater supplies. These groundwater sources are usually contaminated by surface water such as springs and wells.[53] Still, those with groundwater as a drinking water source are at much less risk due to the natural filtering action of soil both above and within the aquifer system.

Researchers in Washington state identified two factors contributing to the highest risk of giardia infection. They were drinking untreated surface water and having more than one sibling between three and ten years old.[54] These factors support the theory that the actual route of the infection is the waterborne acquisition of the cyst by a human, and the subsequent transmission of the infection to others. In other words, the water is the origin but person to person contact is the route of transmission. Reinforcement of this theory was also seen in another Washington state study in which no difference in the prevalence of giardia was found in children that used surface water over those using groundwater.[55]

The specific origin of waterborne giardia in the vast majority of outbreaks remains unknown. Wild and domestic animals as well as man have been implicated as the cause of giardiasis outbreaks. Beavers have been blamed as the source that originally transferred the disease to humans.[56] However, in outbreaks in the states of Washington and New Hampshire it was found that human waste or sanitary violations of individual sewage disposal systems had first infected the beaver.[57] It is believed that the beavers actually acquired giardia from man and only acted as a reservoir for the growth of the disease.

An increased number of giardia outbreaks occur in the summer months to visitors of recreational areas. Either contamination increases at this time, or the supply is always contaminated and the extent of the outbreak is determined by the number of people visiting and using the supply.[58]

Testing for Giardia
Unfortunately, the total coliform test does not detect giardia. In a significant number of outbreaks, minimal or no bacterial contamination was discovered.[59] Therefore, giardia can be found in drinking water that

meets the coliform standard.

Testing water for giardia is pretty much out of the question. Available methods lack efficiency, precision, and sensitivity. They require a large number of samples to perform an accurate measurement.[60] It is also very expensive and the very uncertain results make it not worth the trouble.

Reducing Giardia in Drinking Water

According to the EPA, no correctly operating treatment plant using filtration and disinfection has been blamed for a waterborne giardiasis outbreak.[61] Reported outbreaks of giardiasis have occurred only in water systems where no treatment exists, or where treatment has been inadequate. Diatomaceous earth filtration is considered 100% effective in the removal of giardia cyst.[62] A high percentage of bacteria and turbidity are also removed using finer grades of diatomaceous earth.

Temperatures above 122°F will kill the giardia cyst,[63] so boiling water is a good precautionary move when traveling in the outdoors. At home the chances of a giardia outbreak are slim if the municipal treatment system practices filtration and chlorination. If the source is untreated surface water, filtering in the form of a home treatment device may be the only recourse to ensure giardia-free water. Because giardia lamblia is rather large, 7 microns, compared with other microorganisms, most commercial filters will remove it.

VIRUSES

Certain viruses have also been found to contaminate water supplies. Viruses are transmitted from person to person and multiply only in a living host.[64] Often the cause of unexplained illness, viruses will likely dissipate at their own rate. The majority of healthy people will fully recover from most viruses with little medical expense. Viruses related to waterborne disease are generally **enteric viruses,** which are related to the intestines. Enteric viruses can survive for long periods in waterways.

The extent to which viruses in drinking water cause disease is unknown. All scientists agree that water should be bacteria and virus free. But, like bacteria, the actual health effects of viruses in water are poorly understood. Many viruses cannot easily be traced to water transmission. A person picking up a virus from drinking water may carry it, experiencing no symptoms, yet transmit it to another person who then develops acute symptoms. Some viruses have a long interval between infection and the appearance of illness.[65]

Hepatitis A, Adenovirus, and Norwalk are all viruses for which there have been documented outbreaks. Norwalk virus alone has been found to be the cause of 23% of all waterborne disease outbreaks.[66] The reason only a very few viruses have actually been discovered is due to the limitations of testing and epidemiological studies. For 36% of the waterborne diseases reported between 1971 and 1985 by systems using surface water, no specific biological organisms could be found nor specific cause determined. The EPA thinks many of the unidentifiable causes of these illnesses may be viruses.[67]

Water treatment plants were not designed to remove viruses, and treatment designed to remove bacteria does not mean viruses are also eliminated.[68] Small amounts of viruses that are discharged to rivers and streams are considered a potential health hazard to downstream communities.[69] Although, according to the EPA, properly operating conventional treatment plants utilizing filtration and disinfection can attain 99.99% removal or inactivation of enteric viruses,[70] the effectiveness of virus removal is not known because of the difficulty in monitoring viruses in the water supply.

Virus survival is also variable depending on the water source. Although enteric viruses are more resistant to chlorination than indicator bacteria,[71] the effect of chlorination varies with specific viruses. Disinfection is ineffective if turbidity is high and chlorine does not have enough contact time with the water and viruses.

The level and type of viruses found in a particular water supply will be dependent on the level and type of viruses excreted by the local human population. The ability of viruses to survive in waterways depends on the surrounding natural conditions.[72]

A high percentage of clay in soil is associated with virus removal from runoff moving toward groundwater. Sand and gravel soils do not foster good removal, and of course aid in rapid movement to groundwater. Fractured limestone aquifers and shallow water tables, let water migrate over long distances. It is believed that viruses have the ability to travel more quickly in groundwater than chemical contaminants, and their potential for pollution of subsurface waterways is underestimated.[73] Low pH, acidic soil is a greater aid in adsorption of viruses onto soil particles, whereas soil with a high pH will generally not adsorb them as well. Viruses apparently survive longer at lower temperatures. Pollution, sunlight, suspended solids, and saline conditions, are believed to be detrimental to viruses.[74] Therefore the absence of sunlight, suspended solids, and lower temperatures in groundwater make survival of viruses for long periods very likely.[75]

Testing for Viruses

It is not possible to be certain of an absence of viruses in drinking water by current testing methods. No method exists to cover all viruses in all types of water sources. It is nearly impossible to test for all of the more than 100 types of human viruses found in drinking water. In addition, routine testing methods are unavailable for over half the known viruses in human waste. Add this to the limitations of the methods in existence, and it is unlikely that more than 10% of the viruses known to be in domestic sewage are detectable by current routine testing methods.[76] Science is virtually in the dark ages as far as knowledge of virus detection and the associated health effects are concerned.

Indicator tests, like total coliform, have been suggested as an indicator of viruses because viral disease has rarely been due to drinking water with elevated coliform levels.[77] Viruses have, however, been detected in water meeting turbidity and coliform standards.[78] Consequently, turbidity and total coliform can only be considered indirect indicators of virus contamination.

Total coliform is a good test for pathogenic bacteria, but is not a reliable virus test. A positive coliform test leaves little doubt that viruses are also present, but lack of coliforms does not automatically mean viruses are not there. There are major differences in bacteria and viruses that limit the usefulness of total coliform as an indicator of viruses. For one thing, virus numbers vary in comparison to bacterial counts because viruses are excreted by those infected in irregular numbers.[79] While coliforms are good indicators of fecal contamination, they are much different than viruses in how they survive in the environment or during water treatment. In any case, although indicator tests are only limited in their association with viruses, they will signal dangerous situations like sewage contamination that are characteristic of both bacteria and virus contamination.

Reducing Viruses in Drinking Water

Indentifying and removing sources of biological waste is the best precaution for private wells. Lime soda ash, used in water softening was reported to exceed 99.9% removal efficiency of viruses.[80] According to the World Health Organization, activated carbon is also a good virus remover, but to a lesser extent. They advocate the use of granular activated carbon (GAC) and the development of an alternative disinfection method for virus removal to minimize THM and halogenated synthetic organic chemical (SOC) formation.[81] Generally, however, the need to chlorinate to remove risk of waterborne disease outweighs the risk of SOC production.

LEGIONELLA

The pathogen Legionella pneumophila was a relatively obscure bacteria until 1976 when 34 people died after an outbreak at the annual convention of the Pennsylvania Department of the American Legion in Philadelphia. The acute infection resulting from Legionella (L.) pneumophila, is called legionellosis and is currently associated with two different diseases, Pontiac Fever and Legionnaires' disease. Both are non-communicable.

The Centers for Disease Control estimates that between 50,000 and 100,000 cases of legionellosis occur annually in the U.S., with an unknown amount due to drinking water.[82] Since 1976 a number of deaths from Legionnaires' disease have been reported. Two people in their sixties died in Maryland in June 1986.[83] One employee at a federal office building in New York City (Manhattan) died from Legionnaires' disease.[84] Legionella pneumophila was discovered in the building's water tower and in the air around the desk of the woman who died. Six workers at the **New York Times** developed Legionnaires' disease in June and July 1985.[85] Death from Legionnaires' disease is still rare, however, particularly when compared to some other diseases.

Sources of Legionella in Drinking Water

The only scientifically documented habitat for Legionella pneumophila is damp or water environments. Scientists point out the error of referring to L. pneumophila as a contaminant. It occupies an ecological niche just as hundreds of other microorganisms in the water environment.[86] Currently no outbreak has been directly associated with a natural waterway such as a lake, stream, or pond. Evidently human activity and certain manmade systems like cooling towers, plumbing components, and fixtures harbor or grow the organisms. Therefore, while L. pneumophila may be common to natural water, when taken into distribution systems they can proliferate due to temperature, stagnation, and other factors.[87]

Faucet aerators and shower heads that restrict water flow and cause sludge buildup are chief areas where L. pneumophila can grow to a level that can cause disease.[88] Legionella pneumophila has been discovered in the shower heads, faucets, and hot water tanks in hospitals, hotels, factories, golf course clubhouses, and homes.[89] A survey of a hospital water system showed that legionella pneumophila can exist for a long time within the system.[90] Outbreaks have been reported in hospitals where L. pneumophila was detected in the shower heads and water in the system.

It is believed that shower heads and faucets can emit very small particles called aerosols that harbor L. pneumophila.[91] Aerosols, simply due to their size, can reach the lower respiratory tract of humans.

A link between the presence of L. pneumophila in the water system and Legionnaires' disease in susceptible hospital patients has been established by the medical community.[92] It is this abundance of ill people and the nature of the water system that has resulted in outbreaks in hospitals. The great majority of people who have come down with Legionnaires' disease were immuno-suppressed or compromised due to illness, old age, heavy alcohol consumption, or heavy smoking. Although some healthy people have acquired Legionnaires' disease, outbreaks which have included healthy individuals have usually resulted in Pontiac Fever, the milder type of legionellosis.[93]

The fact that L. pneumophila exists in a water system does not necessarily mean disease is inevitable. Legionella bacteria have been detected in systems where no disease or random unconcentrated cases were found.[94] The condition or susceptibility of the host or patient is believed to be the single most important factor in whether the infection develops in that individual.[95]

Worries over whether Legionella is widespread in city water supplies were partially dispelled in a study by the Pittsburgh Department of Water. While they could not detect legionella pneumophila anywhere in the distribution system, it was found at low numbers in the Allegheny River, the drinking water source for Pittsburgh, and in some hospital tap water samples.[96] They also reported that metals leaching from pipes (zinc, iron, and potassium) can aid in growth of legionella. Low levels of metals in the plumbing system associated with corrosion may be one of the main factors in growth and survival of L. pneumophila in drinking water.

The water and sediment on the bottom of hot water tanks may also be a source of legionella which can contaminate shower heads and plumbing fixtures. Legionella can proliferate at temperatures normally found in the sediment of hot water tanks.[97] Legionella also typically grows in places where water stagnates and sediments accumulate, like hot water tanks.[98] Another study in Pittsburgh concluded that the survival of L. pneumophila in hot water storage tanks is related to the amount of sediment in the tank. The sediment has a nutritional benefit for natural bacteria which in turn promotes the growth of L. pneumophila.[99] The elimination of stagnant hot water could reduce L.pneumophila levels that pose a risk to those with depressed health.[100] While L. pneumophila, if present, will most likely be found in the hot water plumbing system, at least one study reported it in cold water without apparent sediments.[101]

Optimum temperature for L. pneumophila growth is 99°F (89–108°F). Some hospitals and hotels keep the temperature of hot water heaters low to save money. Even the Joint Commission Accreditation of Hospitals has required hot water in showers not be above 120°F to prevent scalding.[102] These lower temperatures, approaching the optimum growth range, could induce the growth of Legionella.

Limited information exists on individual home water systems as a source of legionella. An EPA sampling and survey of 92 domestic hot water heaters found two positive samples taken from electric noncirculating hot water heaters.[103] One was from a municipal water source, the other from an individual shallow dug well. The study concluded that legionella in home water systems is rare. Another study determined that up to 30% of the homes in the U.S. may contain L. pneumphila.[104] The researchers linked two cases of Legionnaires' disease to the home water supply. Both individuals were restricted to their homes due to medical problems.

If legionella enters the hot water tank, it can proliferate and spread throughout the plumbing system including aerators and shower heads. Evidence suggests that Legionnaires' disease is acquired from the air in showers as the water forces the bacteria from the shower head into the air. This is believed to be the main exposure medium, however, ingestion should not be ruled out.

Legionella has the ability to survive conventional water treatment. It is believed to be considerably more resistant to chlorination than coliform bacteria, and can survive for extended periods in water with low chlorine levels. In addition, it can gain access to municipal water systems through broken or corroded piping, water main work, backsiphonage, and cross connections.

Testing for Legionella

While high levels of legionella can be detected analytically, lower levels common to water supplies cannot.[105] The testing methods used today lack the sensitivity to detect low levels of legionella. These levels, however, may prolilferate to higher levels. Common indicators of microbiological contamination of drinking water, coliform or HPC tests, cannot accurately determine the presence of legionella.[106] Even if it were detected in a drinking water environment, it does not mean there is reason for concern, or that it is significant from a public health standpoint.

Reducing Legionella in Drinking Water

Legionella is abundant in ambient water, and because treatment will probably still allow small numbers to pass through, it is possible that they might frequent the plumbing system. Control measures are available and effective in combatting legionella, but they are advantageous only when an outbreak occurs. Raising the temperature to above 130°F in the hot water tank or chlorinating to a level of over 2 ppm of chlorine residual are recommended.[107] However, if an outbreak does not exist, the resulting increase in THMs and chlorinated organics and the possibility of scalding does not warrant using these methods simply as a precaution.

Concern exists mainly for immuno-suppressed, elderly, or ill individuals. Remember that death from Legionnaires' disease, while it has received much media attention, remains very rare. If there is concern however, keeping the hot water tank temperature above 130°F should control legionella organisms. Beware that scalding can occur, and warn guests of the high temperature. Also, remember to super-insulate the tank to save energy. Routine hour-long periods of "heat shock" at 160°F should kill any bacteria that exist in the tank. Periodic draining of the tank to remove sludge that accumulates on the bottom, a breeding ground for legionella and other microorganisms, is a good idea. Be wary of aerators, washers and gaskets in the water system and eliminate dead end plumbing fixtures. Replacement of black rubber washers with ones that were not conducive to growth has also proven effective in legionella control in some cases.[108]

9. RADIOACTIVITY

Eighty-eight chemical elements exist naturally. Their interaction is the basis for life processes and artificial operations initiated by man. Each element is comprised of a specific number of electrons. How a certain element reacts or behaves is based on the number of electrons it has. They surround a central nucleus consisting of an equal number of protons. The attraction between the positively charged protons and negatively charged electrons keeps the atom together.

Because the positively charged protons confined in the small area of the nucleus will repel each other, neutrons are necessary to minimize the effects of the protons which threaten to pull the nucleus apart. The number of neutrons in the nucleus exists independently of the number of electrons. The amount of neutrons is determined only by what is required to keep the protons together. A specific element, then, can have a different number of neutrons but will always have the same number of protons. Radon, for example, has 86 protons and as long as this is so, it will remain radon. Radon, like all other elements can have a varying number of neutrons in its nucleus.

Atoms of the same element with different numbers of neutrons are called *isotopes.* Some isotopes are unstable due to the imbalance in the nucleus and are *radioactive.* Radioactive elements, or *radionuclides,* emit

Table 9-1	
type of radiation	particles emitted
alpha particle	2 protons (positively charge particles) and 2 neutrons (neutral particles)
beta particle	1 electron (negatively charged particle)
gamma ray	electromagnetic radiation

particles to achieve a balance in the nucleus of the atom. By pushing out certain types of particles, the atom becomes more stable. Every radioactive element strives for this stability. The radiation emitted from radionuclides comes in three basic forms, alpha, beta, and gamma radiation (Table 9-1).

The process of pushing out alpha or beta particles, called alpha decay or beta decay, usually results in a new element. The isotope which is decaying is called the **parent** and the resulting isotopes are called the **progeny**. The time it takes for one half of the parent isotope to decay to another element, the **daughter,** is known as the **half-life.** This amount of time varies with each isotope. Some isotopes take as long as billions of years while others as little as a fraction of a second.

Levels of most contaminants in drinking water are expressed in terms of the weight of a contaminant in a liter of water. In considering radioactive levels, the chief consideration is how many particles are discharged or emitted. The number of these particles or rate of particle emission will determine the effect on living systems, like humans. The concern is how **active** the sample is. The unit of measure then, must reflect the activity or number of particles emitted. The half-life of a radionuclide is a function of its activity. The longer the half-life, the lower the activity and vice versa.

The activity of an isotope is measured in curies (Ci). One trillionth of a curie is a picocurie (pCi) which is the unit of measurement commonly used. In drinking-water tests for radionuclides, the results are given in picocuries per liter (pCi/L).

THE OCCURRENCE OF RADIONUCLIDES
IN DRINKING WATER

Natural radioactivity is usually associated with underground water and specific geological conditions. Radioactivity resulting from man's activities is usually confined to surface water. Natural isotopes decay from

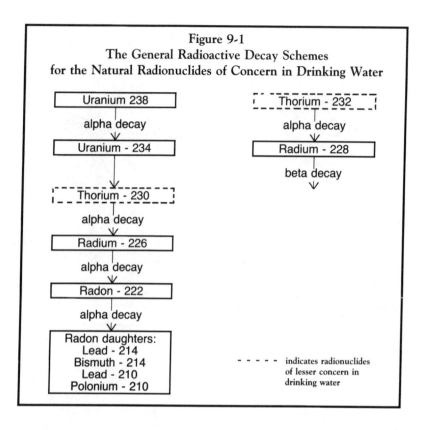

Figure 9-1
The General Radioactive Decay Schemes
for the Natural Radionuclides of Concern in Drinking Water

underground rockbeds and geologic formations into the aquifers flowing through them. The natural radionuclides of most concern due to their occurrence in drinking water and their effects on human health, are radium-226, radium-228, radon-222, and uranium. All of these natural radionuclides are related by natural decay processes (Figure 9-1). With the exception of radium-228, they all decay by alpha particle emission. Radium-228 decays by beta emission. Consequently, sources of natural radionuclides are often similar, whether they are from natural geologic formations or become a pollutant through man's activities. According to the EPA, however, radon, uranium, and radium are rarely found together at high levels.[1]

High levels of natural radionuclides are generally associated with certain sections of the country. The geology of an area carries considerable weight as to the likelihood of natural radionuclides in drinking water. Figure 9-2 outlines the general areas in the United States where each

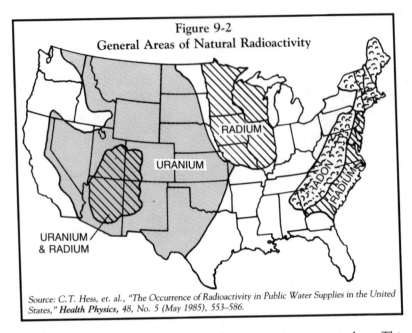

**Figure 9-2
General Areas of Natural Radioactivity**

RADIUM

URANIUM

RADON

RADIUM

URANIUM
& RADIUM

Source: C.T. Hess, et. al., "The Occurrence of Radioactivity in Public Water Supplies in the United States," **Health Physics**, 48, No. 5 (May 1985), 553–586.

natural radionuclide common to drinking water is most prevalent. This must only be interpreted as a *general* guide because geology can be region specific, common to a very wide area, but quite variable within some smaller areas.[2] Generally, the highest levels of natural radionuclides are found in aquifers with low yield or small water systems, like private wells.[3] Consequently, exposure would be more significant in these water supplies compared to larger public water supplies serving several hundred people.

Uranium

Uranium has been found in both surface water and groundwater,[4] but concentrations in groundwater are usually higher.[5] Granite metamorphic rocks, lignites, monzonite sand, and phosphate deposits commonly contain uranium. The highest levels of uranium in water are found in the the west: the Colorado Plateau; Western Central Platform; Basin and Range; and Rocky Mountain system (Figures 9-2 and 9-3).

Natural underground rock formations which contact groundwater, uranium mine tailings, phosphate ore mining deposits, and fertilizers are all sources of uranium. Phosphate mining and processing of phosphoric acid disrupts the equilibrium existing between uranium and radium-226 in the ore matrix. Uranium follows the phosphoric acid end product and

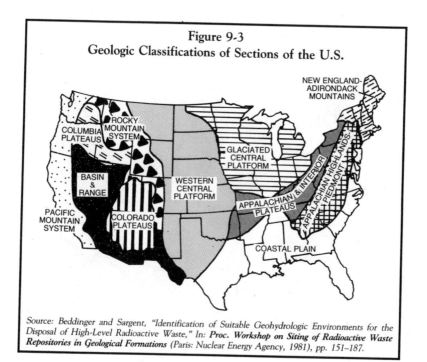

Figure 9-3
Geologic Classifications of Sections of the U.S.

Source: Beddinger and Sargent, "Identification of Suitable Geohydrologic Environments for the Disposal of High-Level Radioactive Waste," In: **Proc. Workshop on Siting of Radioactive Waste Repositories in Geological Formations** *(Paris: Nuclear Energy Agency, 1981), pp. 151–187.*

the resulting fertilizers, while most of the radium (and daughter isotopes like radon) are found in the gypsum waste.[6]

Estimates by leading researchers are that 2,500 to 5,000 community water supplies in the United States contain more than 5 pCi/L of uranium.[7] About 300,000 people on public groundwater systems receive water with over 10 pCi/L of uranium.[8] Sections of Colorado, Idaho, Montana, and a large part of California have a high potential for elevated uranium levels in water because of the presence of granite rocks.[9]

As is the case with the other natural radionuclides, uranium levels are highest in private wells. Though uranium concentrations in the eastern section of the country are typically quite low,[10] levels in many private wells have been measured at over 40 pCi/L. Concentrations have been found in the hundreds of pCi/L in wells in Connecticut, New Hampshire, and Georgia, and levels in the thousands of pCi/L in Maine and Colorado.[11]

Radium

Radium is found predominantly in groundwater. The areas of the United States where radium is detected most frequently are the Coastal

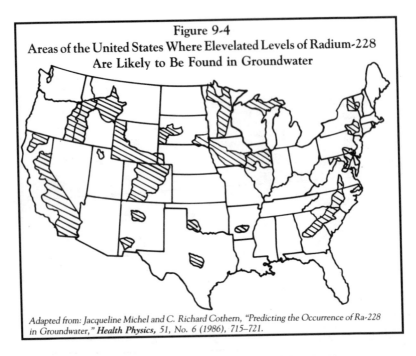

Figure 9-4
Areas of the United States Where Eleveleted Levels of Radium-228 Are Likely to Be Found in Groundwater

*Adapted from: Jacqueline Michel and C. Richard Cothern, "Predicting the Occurrence of Ra-228 in Groundwater," **Health Physics,** 51, No. 6 (1986), 715–721.*

Plain, Piedmont, north Central Platform, and Colorado Plateau regions (Figure 9-2 and 9-3).[12] Large areas of the northern midwest encompassing certain areas of Minnesota, Wisconsin, Iowa, Illinois, and Missouri have granite rocks or sandstone aquifers with elevated Total Dissolved Solids (TDS) levels. This area is considered to have the highest reported radium levels in the United States.

Most of the information on radium levels in drinking water is based on surveys of public water supplies where concentrations rarely exceed 50 pCi/L. But radium is essentially a problem for small groundwater systems, like individual wells. Therefore, based on the levels found in public water supplies, certain areas where radium is not expected may have private wells with elevated radium levels.[13]

The occurrence of radium-228 in groundwater is also dependent on the geology of the aquifer. Granite rock aquifers, sandstone aquifers containing feldspar, and quartoze sandstone aquifers with high TDS are considered to have the highest probability of having groundwater with increased radium-228 levels.[14] The regions of the country where high radium-228 in groundwater is likely to be found are outlined in Figure 9-4.

Radon

Unlike other radionuclides of concern in drinking water, radon is a gas. This colorless and odorless gas is a serious potential problem in indoor air. Indoors, radon or radon daughters will attach to air particles like dust or cigarette smoke, which when inhaled will attach to the walls of the lung. Health risk due to radon is higher in homes of smokers.

Radon enters homes generally through open foundations such as dirt floor basements, cracks or holes in the basement or foundation, crawl spaces, sump openings, and from certain building materials themselves (Figure 9-5). It can filter up through the soil from bedrock and build up inside a home to dangerous levels. Simply the disturbance of the soil during construction could contribute to higher levels in the house when it is finished. Location of the home and the geology of the earth below it are the dominant factors in determining radon levels in the home and in drinking water. [15] Other considerations like ventilation and construction materials are less critical by comparison.

At first, airtight energy-efficient homes, where fresh air exchange is minimal, were discovered to have high radon levels. But subsequently, older homes with drafts, thought to be generally immune from the buildup of radon, were also found to contain high levels. Consequently, the factor influencing radon levels in homes may not necessarily be the rate of air turnover or exchange, but the rate at which the gas enters or infiltrates the home. The difference in air pressure between the outside and the inside of the house is believed to force the radon inside. This difference in pressure is due to outside wind currents and higher indoor temperatures.

One must keep in mind that radon is a byproduct of nature. The role of radon in the geologic decay process is perfectly natural. Unconfined in the outdoors, the gas dissipates and will not concentrate to high levels. It is when one decides to build a house and set up an artificial atmosphere that radon is allowed to accumulate to dangerous levels.

Drinking water is often overlooked as a source of radon. Radium decay to radon in geologic formations underground can result in the diffusion of the gas into groundwater. Radon, because it is a gas, will desorb from water easily into the air during normal home activities like running water, taking showers, flushing toilets, and washing clothes and dishes. Efficient vacuum pumping in a well system allows no release or escape for the waterborne radon except into the house. Researchers in Maine found a significant correlation between radon in water supplies and airborne radon levels inside the home. [16] The contribution of water to total airborne radon will depend, of course, on the amount of radon in the water and

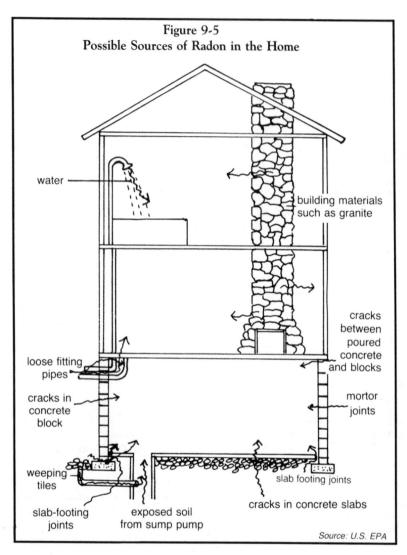

Figure 9-5
Possible Sources of Radon in the Home

water

building materials
such as granite

cracks
between
poured
concrete
and blocks

loose fitting
pipes

mortor
joints

cracks in
concrete
block

weeping
tiles

slab footing joints

slab-footing
joints

exposed soil
from sump pump

cracks in concrete slabs

Source: U.S. EPA

the level of the other combined sources.

Radon in drinking water is not currently considered one of the major sources of radon in indoor air. However, radon from the tap can contribute over 90% to the total indoor level of radon if the groundwater has high radon levels and the direct emission from subsurface geologic formations into the home is low.[17] One study determined the 1,124,000 pCi/L of radon found in a water supply accounted for almost 100% of the 2,000 pCi/L found in the indoor air.[18] It is generally accepted that a radon level of 10,000 pCi/L in water converts to about 1 pCi/L of radon in the air.[19] In any case, the risk from airborne radon is considered serious enough that the EPA is considering a Maximum Contaminant Level (MCL) for radon in drinking water of between 200 and 2,000 pCi/L.[20]

The smaller the water distribution system the greater the chance for high concentrations of radon due to its short half-life. In high radon areas, private wells are likely to have more radon than community systems.[21] This is probably the result of the low capacity of the wells which allows little time for the decay of radon within the system. In addition, there is almost no aeration in smaller systems compared with larger, public water systems. Radon in private wells is a serious problem in some areas. Levels as high as 2,000,000 pCi/L have been detected in some cases.[22] The highest reading from a monitoring survey of 200 drinking water wells in Connecticut was 130,000 pCi.[23] One well in Maine had a concentration of 760,000 pCi/L.[24]

The occurrence of radon cannot be specifically predicted due to the uniqueness of each groundwater situation. However, granite rocks and the tailings from phosphate fertilizer processing and uranium mines are chief sources of high activity.[25] Mine tailings contain large levels of radium and have been known to be used as fill at building sites. When radium decays, it can cause high radon levels in homes built on these locations. In Grand Junction, Colorado and Cottonwood, South Dakota radioactive mill tailings were used as fill during the construction of some homes. An increase in congenital birth defects occurred in babies born to mothers residing in these homes, and an elevated incidence of leukemia was reported in Grand Junction.[26]

The highest radon levels found in groundwater have been in mountain states, particularly New England, the Appalachian Highlands and Piedmont regions including the states of Pennsylvania, Virginia, Maine, New Hampshire, Vermont, Rhode Island, and Massachusetts.[27] The Rocky Mountain and Pacific Mountain states are the next highest, and the midwest is the lowest. The main factor in determining radon levels is the rock type in the aquifer. Radon levels in groundwater are generally highest in areas where granite rocks exist.[28]

HEALTH EFFECTS OF RADIONUCLIDES IN DRINKING WATER

Those doing any further reading on the health effects of radioactivity in drinking water will want to be aware of yet another unit of measure called the REM. The REM is a measure of the effect of an isotope's activity on humans. Acute health effects in humans are not normally observed until a level of 10 REMS of total dose equivalent is reached. However, just as important are the long term effects of low doses of radiation which are poorly understood. Scientists are concerned that no radiation level may be safe. But low background radiation exists everywhere. As a matter of fact, the National Council on Radiation Protection and Measurements raised the estimate of U.S. radiation exposure from 170 to 360 millirems yearly.[29] The increase is primarily due to radon gas seeping into homes, which accounts for 55% of the total radiation exposure.

The most serious health effects associated with radiation exposure are birth defects and cancers. The fetus is very vulnerable to radiation damage and ionizing radiation has been clearly implicated as carcinogenic. Known health effects of radioactivity include:

- developmental and teratogenic effects including abnormal skeletal and central nervous system development;
- genetic effects, including mutagenic changes in DNA, hereditary effects, and diseases as a result of mutations;
- somatic effects including cancer and leukemia, shortening of life span, cataracts, testicular damage, and sterility;[30] and
- kidney damage from uranium.[31]

Considering the numerous other sources of radiation, data cannot statistically prove a relationship between radionuclides in drinking water and cancer. Generally, low levels of radionuclides are found in drinking waters, making it difficult or impossible to accurately measure the existence or extent of any health effects. Even so, there is evidence that radionuclides in drinking water are capable of causing cancer and genetic effects.[32] According to Dr. William L. Lappenbusch, a nationally recognized drinking water toxicologist:

> "radionuclides in drinking water cause more cancer than any other stressor in that medium, challenged only by pesticides."[33]

Health Effects of Radon in Drinking Water

Discovery of radon levels of up to 2,700 pCi/ in the air in a Boyerstown, Pennsylvania home led to the identification of the infamous Reading Prong, a geologic formation that stretches from Reading, PA through New Jersey and into New York state. Subsequently, radon has been found in homes across the United States, making it a priority health concern.

Numerous studies have determined a direct association between radon exposure and lung cancer in humans. Unlike many other pollutants, a clear link between elevated levels of radon and lung cancer does exist. The EPA estimates that radon in indoor air causes between 5,000 and 20,000 lung cancer deaths yearly. It is likely the most significant cause of lung cancer among nonsmokers.[34]

There is increasing concern about radon because of its widespread occurrence and significant risk to the population, compared with other pollutants. Americans spend an average of 80% to 90% of their time indoors. The sick, the elderly, and the very young spend almost all their time indoors.

Ingestion of radon in drinking water is also a source of exposure. However, it is one that has not been adequately explored. Researchers in Maine estimate that a person living 60 years in a house which has a level of 20,000 pCi/L of radon in the drinking water would have a one chance in 500 of dying of stomach cancer due to the radon.[35] The chances of acquiring stomach cancer would of course predictably increase with increased levels in the water supply.

According to C. Richard Cothern, Ph.D., a physicist with the U.S. EPA, high levels of radon in drinking water in the U.S. are widespread and the potential risk over lifetime exposure is significant.[36] He estimates that over a 70 year time-span, an average lifetime, 2,000–40,000 incidents of lung cancer result from radon in drinking water. In rural areas where people live most of their lives in the same area and draw their water from private wells, radon levels could be even more significant.

Man-made Radionuclides

As a result of human technological endeavors, man has created over 800 radionuclides. About 200 of these are believed to be potential drinking water contaminants and studies have identified them as carcinogenic.[37] Man-made radionuclides are, of course, a product of the nuclear industry. Generally, nuclear facilities are located near cities and could pose a threat to surface water supplies. Sources of man-made radionuclides in drinking water include: nuclear materials processing plants; nuclear weapons testing; nuclear power plant discharge; medical waste; radionuclide waste depository leakage; nuclear waste compactors; and nuclear accidents.

The chance of contamination of a private well with man-made radionuclides is believed to be slim. Groundwater is considered inaccessible to land deposition from fallout and groundwater stays underground for very long periods of time. Radionuclides with short half-lives will decay before the contaminated water reaches the well. Common man-made radionuclides with long half-lives like Cesium-137 and Strontium-90 bind heavily to soil preventing them from reaching groundwater.[38] Cesium-137 and Strontium-90 are most likely to be found in surface waters because they wash out easily in runoff water.[39] According to generally available information, few man-made radionuclides have been found in drinking water *above* the level of 50 pCi/L.[40] Strontium-90 and Tritium are of the greatest concern below 50 pCi/L.[41]

In any case, the potential for increased levels of man-made radionuclides in water does exist, particularly near facilities that create, use, store, or dispose of them. Documented cased include:

- A General Accounting Office (GAO) survey of Department of Energy (DOE) facilities found Strontium-90 at over 400 times the drinking water standard in groundwater near the N-reactor in the State of Washington.[42] Tritium and/or Iodine-129 were discovered in the groundwater at three other DOE facilities.[43Z]
- Eight out of nine DOE nuclear processing facilities reviewed in 1986 by the GAO were found to have groundwater contaminated with chemical and radioactive waste.[44] In some cases the contaminants have moved off the site or into rivers.
- Release of man-made isotopes from nuclear power plants in Pennsylvania resulted in contamination of Susquehanna River sediments.[45]
- Between 530,000 and one million pounds of uranium dust was emitted into the air from the uranium processing plant in Fernald, Ohio since it opened in the early 1950s. As a result of careless on-site dumping, handling, and storage of millions of pounds of radioactive wastes, organic solvents, and lead wastes, Fernald is one of the three largest nuclear waste dumps in the country.[46] Groundwater and wells in the area are contaminated with elevated levels of uranium.[47]
- Problems with DOE nuclear plants in Hanford, Washington; Rocky Flats, Idaho; and Savannah River, South Carolina have resulted in their shutdown.[48]
- In 1979, one hundred million gallons of radioactive liquid and 1,100 tons of radioactive tailings spilled into the Rio Puerco river, a drinking water source to a number of communities, when a uranium tailing dam broke in Church Rock, New Mexico.[49]
- Inactive DOE waste sites reportedly number 1,447, but the GAO believes the site inventory is incomplete.[50]

TESTING FOR RADIONUCLIDES

Having water tested for radionuclides may prove to be more difficult than other tests. Not many labs actually perform testing for radioactivity. However, many of those that do not are willing to forward water samples, for a small fee, to a lab that specializes in radionuclide testing. But, beware of labs that charge as much as quadruple the test price to unsuspecting clients.

Since there are hundreds of man-made radionuclides and only a few have proven to be a potential problem in water supplies, it is economically unrealistic and unnecessary to test for all of them. Analysis of radioactivity in drinking water is most often done using screening methods to measure the total amount of alpha and beta activity in the water sample. The tests are called, appropriately, **gross alpha activity** and **gross beta activity.** Because most natural radionuclides of concern in drinking water emit alpha particles, and most man-made radionuclides emit beta particles and gamma radiation,[51] one can determine which test to choose based on which group is of most concern. If these screening tests reveal radioactivity in the water supply, testing can be done for individual radionuclides. Methods exist for specific radionuclides that have proven to be potential water contaminants.

According to the federal interim drinking water regulations, monitoring for natural radioactivity in public water systems is required. If gross alpha activity is more than 5 pCi/L for a water sample, then radium-226 must be analyzed. The gross alpha activity is most likely to be due to radium-226, unless the area has high uranium levels in groundwater. If radium-226 is found to be more than 3 pCi/L, then radium-228 must be analyzed. If the Maximum Contaminant Level (MCL) for gross alpha activity, 15 pCi/L, is exceeded, then uranium and radium testing is required.[52]

Sometimes the results of uranium testing will be expressed in ppm rather than in terms of activity (pCi/L). This method of determining the concentration of uranium is less costly than measurements for radioactivity.[53] Be aware that converting the mass measurement (ppm) to activity (pCi/L) may not account for all of the activity of the different uranium isotopes.

Radium-228 was incorporated into the gross alpha activity screen so large numbers of samples would not have to analyzed for radium-228 because of its lengthy and costly testing procedure. It was thought that radium-226 and radium-228 concentrations in groundwater were related to each other. However, several studies have seriously questioned this relationship finding radium-226 and radium-228 ratios were not always consistent as assumed in the grosss alpha activity screen. The concentration of radium-228 was found to be unrelated to radium-226 in certain aquifers.[54] Consequently, some supplies may have levels of radium-228 higher than radium-226, a scenario that was previously not believed likely. As a result, it has been estimated that between 40% and 50% of excess radium-228 levels are not detected.[55] Therefore, gross alpha activity screen cannot consistently determine the total radium concentration unless the ratio of radium-226/radium-228 can be estimated for the aquifer being tested. In areas where it is believed radium-228 may be present, the EPA recommends that states make testing for radium-226 and/or radium-228 mandatory for municipal water systems when gross alpha particle activity is greater than 2 pCi/L.[56]

Only surface water systems serving more than 100,000 people are required to monitor for man-made radioactivity. If gross beta activity is more than 50 pCi/L, extensive analysis must be done to determine which isotopes are present. Two relatively common man-made radionuclides can be found at levels *below* 50 pCi/L, tritium and strontium. They must be measured if the gross beta screen is below 50 pCi/L.[57] Those guidelines are also useful to the individual.

Testing for Radon in Drinking Water

For homeowners or potential homeowners, it is certainly advisable to test for radon in indoor air. Testing air in the home for radon is easy and inexpensive. Because the concern about radon in well water revolves around its contribution to levels in indoor air, testing the air is the best place to start. If levels in the air are found to be high and a private well is the water source, one may want to go ahead and test the water for radon. It would also be wise to consult the state radiation protection office before making any costly radon reduction or mitigation work.

Companies that test radon in air are found everywhere. Some department stores even sell kits. Although the EPA has recently initiated stricter standards for radon testing companies, an evaluation of the efficiency of radon home testing kits can be obtained by sending $2.00 to: Public Citizen Radon Guide, P.O. Box 33487, Washington, D.C. 20033-0487.

The cost is commonly around $10 for a charcoal canister type radon detector. The test consists of placing the canister or canisters of activated carbon in living areas of the home or the basement, where levels are believed to be the highest. Radon and its daughters are adsorbed onto the carbon over time. The canister is then sent to a central lab for analysis.

More accurate measurements can be made using an alpha-track detector which measures over a longer perioid of time. The cost is around $50. It is advisable to test as soon as possible, but also test at different times of the year. Researchers at Pennsylvania State University measured the highest levels of radon in soil during the summer months.[58] The results of these tests would generally determine the plan of action. The EPA gives the following recommendations:

Test Result: 4 pCi/L of air
Chances of dying of lung cancer from indoor radon to a person living in a house for 60 years is estimated at 1 chance in 100.[59]
Action: Some risk, but making any reduction at this low level would be difficult or impossible to achieve.

Test Result: 4 pCi/L to 20pCi/L
Action: Exposure is above average. Action should be taken within a few years.

Test Result: 20 pCi/L to 200 pCi/L
Action: Exposure is greatly above average and action should be taken in the next few months.

Test Result: above 200 pCi/l
Action: Exposure is among the highest observed in homes and action should be taken in several weeks. Consult with local or state radiation officials about possible relocation.[60]

A ten state EPA survey found that one-fifth of the homes had radon levels over 4 pCi/L, equivalent to over 200 chest x-rays per year, indicating that radon could be a problem virtually anywhere.[61] Those living in the northeast are not the only ones who should be concerned. Certain areas of almost every state have the potential for elevated radon levels. Although geology can indicate possible high radon areas, there is no way to know the levels in individual homes without testing. Levels can vary from house to house.

As with any other contaminant, the degree of concern regarding the health risk of radon is up to the individual. Clearly the level of indoor radon will influence the level of concern and the decision about what action to take, if any. High levels like 100 pCi/L in indoor air are certainly a major health risk. While the EPA is concerned that no level of exposure to radon is safe,[62] scientists in Maine point out that worry or anguish over very low levels of radon could affect some people more than the radon itself. In the vast majority of situations there is no cause for urgent action.[63] It generally takes decades of exposure to bring about measureable risk of disease.

If testing of indoor air reveals elevated radon levels, it may be a good idea to test the water to determine what contribution it may be making to airborne radon levels. State Public Health laboratories in some states with high water radon levels may offer radon water testing. The state of Maine, for example, offers this service for a fee to Maine residents. They will send a kit with the required sample bottle and instructions for sampling. Results will be mailed back to customers. As with other radioactive contaminants, some commercial labs will also perform radon testing. Also, an alpha track detector can be placed inside the toilet tank to measure the amount of radon released during flushing.

SAMPLING FOR RADIONUCLIDES

For gross alpha and gross beta activity, sample preservation is not required, but it is advisable to add nitric acid to lower the pH. Samples without a preservative should arrive at the lab within 5 days.[64] For specific radionuclide testing in drinking water, contact the lab for specific instructions on sampling containers and preservation requirements, since they differ for each particular radionuclide. The preservative that is best for one radionuclide could lead to losses in the measurement of another.

According to the EPA, an accurate sampling for radon can be achieved using a small vial. Remove the faucet aerator, allow the water to run for ten minutes, and fill the bottle slowly. Cap the sample quickly and be sure no air bubbles are in the sample.[65] A special sampling bottle with an airtight cap should be obtained from the lab. Send the sample to the lab promptly.

HOME TREATMENT METHODS
Radium

Among the effective treatment methods currently available for removing radium from drinking water are water softening and reverse osmosis. The ion exchange softener is a very common home treatment unit used in hard water areas. A zeolite resin exchanges sodium for metals like calcium and magnesium using ordinary salt. Since it is similar chemically to calcium and magnesium, radium will also be removed if it is present. Fortunately, radium removal will continue after the softening capacity (calcium and magnesium removal capability) of the resin is exhausted. Consequently, maintaining low hardness should indicate good radium removal. Beware of the disadvantages of softening, high sodium content, high corrosivity of the water, and loss of valuable nutrients (See chapter 6). A study of low-pressure, reverse osmosis (RO) units found them over 90% effective for radium removal.[66] Granular Activated Carbon (GAC), common in many home water treatment filters, is ineffective for the removal of radium.[67]

Uranium

Methods of uranium removal from drinking water are only in the research and development stage. Lack of a drinking water standard has minimized efforts in this direction. Treatment methods are confined to the industrial scale such as recovery from mine processes and tailings. Studies have shown, however, that softening and RO have the potential for uranium reduction in drinking water. RO was reported to be 99% effective.[68] GAC apparantly does an excellent job of trapping many of the isotopes in the uranium decay series, with the exception of radium.[69]

Radon

Aerators are the preferred method of radon treatment. This process exposes the water to air which allows the radon gas to dissipate from the water. Aeration methods are being developed for Point-of-Entry (POE) treatment, but are not currently widely available for home use. They are expensive, from $2,500 to $4,400, and a certain amount of maintenance and energy is necessary for their operation.

Granular activated carbon has also proven effective in radon removal from drinking water. Jerry D. Lowry, Associate Professor of Civil Engineering at the University of Maine at Orono, and a leading researcher in radon removal from drinking water, reported removal efficiencies of greater than 99%.[70] GAC units are widely available, fairly inexpensive, and require little maintenance, but have a significant drawback.[71] Decay within the bed occurs due to radon's short half-life. Concerns about the resulting buildup of radon daughters within the unit and the possibility of other forms of radiation being emitted into the air have not been widely addressed. A few radon daughters emit gamma radiation, so those electing to utilize GAC to remove radon from drinking water should either shield the unit with concrete or lead or place it outside the home.[72] Also, the problem of the disposal of the treatment material containing the radionuclides is a consideration. Special disposal methods may be necessary or even required.

Cost estimates by the EPA for GAC Point-of-Entry devices which will effectively remove radon from drinking water run between $650 and $1,200.[73] The price of the activated carbon filter unit will depend on its efficiency. A home POE device with a sediment filter capable of over 99% removal would be priced at about $1,200. Yearly operating costs are about $20.[74] The most effective volume of GAC for POE removal of radon is between one and three cubic feet.[75] A sediment prefilter should be used to remove particles that can clog the GAC bed. Do not backwash the unit to remove particles. This stirs up the carbon and reduces the radon removal efficiency of the GAC.[76]

Man-made Radionuclides

Because man-made radionuclides are very rare in groundwater, there is no reason to treat private wells for them under normal circumstances. Little research has been done on treatment of surface water for the removal of man-made radionuclides. However, traditional surface water treatment methods like filtration and coagulation are considered effective for many of the man-made isotopes.[77] Research has shown that RO removes over 90% of a variety of man-made radionuclides including cesium and strontium.[78]

10. NITRATES

"As water spilt on the ground which cannot be gathered up again. "
— Samuel 14:14

Nitrogen is one of the earth's basic natural elements. In water, nitrogen can occur in the *ammonia, nitrate* or *nitrite* form. Nitrate is the form of most concern and, unfortunately, it is also the most stable. It resists changing to the other forms, but the other forms readily change to nitrate (Figure 10-1). Therefore, most nitrogen in natural water systems easily converts to nitrate.[1]

Sources of Nitrates in Drinking Water

Any nitrogen source should also be considered a potential origin of nitrate contamination. Major sources of nitrogen in drinking water are septic tank effluents, fertilizers, sewage, and animal feedlot waste found in agricultural areas.

Commercial fertilizer, an ammonia and nitrate combination, is a major contributor to nitrate levels in surface and groundwater. Poor farming methods that lead to errosion and soil degradation caused farmers to turn to heavy fertilizer usage in an attempt to maintain crop yield. Between 1960 and 1985 total nitrogen consumption in the U.S. increased almost four-fold.[2] In 1930, fertilizer added to soil in farming areas averaged 8 pounds of nitrate per acre. In 1980, when fertilizer production was at its

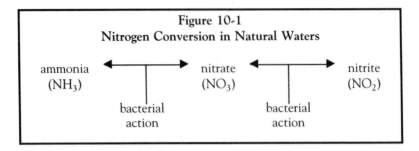

Figure 10-1
Nitrogen Conversion in Natural Waters

ammonia   nitrate   nitrite
(NH_3) (NO_3) (NO_2)

bacterial   bacterial
action   action

highest level, an average of 283 pounds of nitrate per acre was added to the soil.[3]

Fertilizer companies have even been accused of recommending wasteful nitrogen application practices. Research at the University of Nebraska showed that effective usage should cost only $23 per acre for irrigated corn fertilizers, while the heavy usage practices recommended by some company labs resulted in costs ranging from $43–$61 per acre.[4] Over-application of fertilizer leads to build up of nitrate in the soil. Plants cannot use the nitrogen fast enough and nitrates are washed into groundwater and surface water sources.

As a result of nitrogen runoff, many areas of agricultural states like Nebraska, Iowa, and Wisconsin have drinking water with high nitrate levels. A study in Nebraska found over 70% of wells tested had over 10 ppm nitrate-nitrogen.[5] Forty percent of the wells tested in a Wisconsin study exceeded 10 ppm nitrate-nitrogen.[6]

Agriculture, however, is not the sole source of nitrogen. Fertilizer applications to lawns and golf courses account for a large portion of the total nitrate going to groundwater recharge. Septic tanks are also a significant source of nitrate found in groundwater.

Several factors can influence the potential for nitrate contamination of drinking water wells. Some studies indicate that well depth is a determining factor. The shallower the well (less than 50 feet), the greater the chance for nitrate contamination. A study in Illinois found 81% of dug (shallower) wells, and 34% of drilled (usually deeper) wells contained more than 10 ppm of nitrate-nitrogen.[7] This is not surprising since shallower wells are more likely candidates for contamination of many different pollutants, not just nitrate. In a Wisconsin study, however, well depth was *not* found to be a consideration in predicting high nitrate levels. Depth of the well casing was determined to be more of a factor in preventing nitrates and bacterial contamination than well depth.[8] Levels of nitrate in groundwater will often be higher after rainy periods and lower during dry periods as nitrogen is washed from the surface into aquifers.

Health Effects of Nitrates in Drinking Water

Concern about nitrate in drinking water stems from the fact that it reduces to *nitrite* in the gastro-intestinal tract in humans. There nitrite combines with hemoglobin, which is responsible for carrying oxygen from the lungs to the bloodstream. It forms a new compound, **methemoglobin,** which cannot transport oxygen. A victim of **methemoglobinemia** could theoretically suffocate to death.

Nitrate in drinking water is typically not a problem to the adult population. Infants younger than three months old, are the most susceptible to methemoglobinemia. There are several reasons for this:

1. hemoglobin is underdeveloped and in a vulnerable form in the infant, making methemoglobin formation much more likely;
2. the enzyme necessary for preventing methemoglobin formation is not developed at birth;
3. the infant's digestive tract cannot secrete gastric acid, allowing bacteria to live in the stomach and convert nitrate to nitrite; and
4. in an adult, only approximately 10 percent of the nitrate ingested is converted to nitrite, while in the infant, 100% of the nitrate consumed is converted to nitrite.

Food is believed to be the highest source of nitrate.Most common vegetables like beets, celery, lettuce, and spinach contain high levels of nitrates, as do meat additives. However, in locales where the nitrate-nitrogen level in water approaches 100 ppm, up to 70% of the total nitrate/nitrite intake can come from water.[9] The combination of nitrate from food and nitrate from water could be a problem for infants.

A review of infant methemoglobinemia studies in 1951 found none of the cases reported were associated with water containing nitrate below 10 ppm nitrate-nitrogen.[10] This seems to be the no-effect level. However, according to the National Academy of Science (NAS), there is no guaranteed safety margin in that level.[11]

At least seven Iowa infants have died of methemoglobinemia since 1945 when the disease was first recognized.[12] A 1950 study of Minnesota methemoglobinemia records found 14 deaths in 139 cases believed caused by water from farm wells.[13] No breast-fed infants developed the disease.

Many infants have been known to drink water with nitrate-nitrogen levels greater than 10 ppm, the current Maximum Contaminant Level, without getting methemoglobinemia. Two scientists pointed out the rather surprising fact that certain wells in California have nitrate levels in excess of 2,000 ppm without any reported cases of methemoglobinemia.[14] Another pair of researchers even tried to induce methemoglobinemia in infants by giving them sterile well water with nitrate. They used a dose that would be the same as using well water with 2,000 ppm nitrate comprising half of the baby formula.[15] This unsuccessful attempt, with levels far above those normally found in positive cases of methemoglobinemia, leads investigators to believe that while nitrate is a contributing factor, another factor or factors must be present in order for methemoglobinemia to occur from ingestion of water with low nitrate levels.

In one study, researchers found increased methemoglobin levels in infants with diarrhea, and concluded that bacterial contamination of water and baby formula could be one reason.[16] The majority of methemoglobinemia cases reported in the world health and scientific literature are related to private and/or bacterially contaminated wells. Shallow wells in farm areas are prime candidates for bacterial contamination associated with diarrhea. All things being equal, water from a shallow, private well is more likely to induce methemoglobinemia than water obtained from a deeper municipal well with the same nitrate level, because the deep municipal well is far less likely to be bacterially contaminated.

Methemeglobinemia is believed to be quite rare today, particularly in countries like the United States with comparatively high health standards. This does not mean that there is no reason for concern. For one thing, methemeglobinemia is not a reported disease in the United States. Consequently, an accurate assessment of the incidence of the disease does not exist. Elevated nitrate levels are routinely reported in many areas, and one infant died from ingesting formula made with water from a private well contaminated with nitrate in 1986.[17] The two-month old died after ingestion of water with 150 ppm of nitrate from the family well. The well was only 30 feet deep, close to cropland, and less than 40 feet from the septic tank. The infant may have been saved, but the symptoms of methemoglobinemia were not recognized by the hospital.

The process of nitrate conversion to nitrite by bacteria in the stomach is the determinant factor in other possible health effects as well. Nitrites react with normal stomach fluid to form highly carcinogenic nitrosamines. There are implications of upper digestive tract cancers in humans from high nitrate consumption, but studies are not considered conclusive by most of the scientific community.[18] Many however, see a clear reason for concern and a need for continued assessment of nitrate effects.

Iowa farmers have an increased risk of stomach cancer, particularly in counties with high corn, milk, and cattle production.[19] Marty Strange and Liz Krupicka of the Center for Rural Affairs, in a review of several studies of leukemia incidence in farming areas found nitrate in drinking water as a possible common denominator.[20] In the central Platte River valley of Nebraska, known as the "leukemia belt," nitrate levels are particularly high.

A higher incidence of birth defects may also be associated with nitrate. Australian researchers found a significant increase in birth defects when mothers drank groundwater containing even low levels of nitrate, 5 to 15 ppm nitrate-nitrogen.[21]

Sampling for Nitrates & Nitrogen Testing

Sulfuric acid is needed as a preservative, unless the sample is being sent through the mail. In any case, put the sample on ice and get the sample to the lab within 24 hours.

Testing for Nitrates & Nitrogen

Several different testing methods are used to analyze for nitrate, but none are regarded as particularly precise or sensitive. All methods have a reproducibility problem.[22] Reproducibility — being able to obtain close to the same result if the test is repeated — while a consideration in all testing, appears to be an inherent problem in nitrate testing. The results of nitrate analysis could be off by 1 ppm based solely on the method.[23] However, this does not dismiss nitrate testing as a waste of time. Knowing the potential limitations of a test can only enhance the interpretation of the results.

Test the water routinely because nitrate levels may fluctuate throughout the year. Levels will vary over time as the aquifer is recharged by runoff water. It would be advisable to test at least twice per year, once in the spring when the aquifer is high and once in the fall when it is low.

Test Results and Maximum Contaminant Level (MCL)

Currently, an MCL of 10 ppm for nitrate-nitrogen exists based on its possible toxicity and fatal poisonings to infants.

It should be understood that the nitrate results can be reported in two different ways. Often the level is expressed as nitrogen, the elemental component of the nitrate ion, as is done in this chapter. However, sometimes levels are expressed simply as nitrate. If this is the case, the MCL would be 45 ppm. So:

$$45 \text{ ppm nitrate } (NO_3) = 10 \text{ ppm nitrate-nitrogen } (NO_3\text{-}N)$$

Sometimes this may be confusing as it is not always clear which is being used. If there is uncertainty, be sure to question the result.

Reducing Nitrates in Drinking Water

For those concerned about methemoglobinemia, all that appears necessary is an awareness of the problem and a limiting of nitrate exposure to the infant. This may include testing the water source for nitrate, particularly if it is a private well and it is used to draw water for the child's formula. Methemoglobinemia has not been a problem in breast-fed infants.

Since nitrate levels are not significantly reduced by current conventional water treatments, those using community water should be aware of nitrate also. Monitoring for nitrate may not be required, depending on the size of the municipal system, or it may be performed only once per year. Therefore, it would not indicate a seasonal change in concentration.

The use of bottled water is an advisable alternative if one has an infant or are expecting a child. Bottled water is priced generally under $2 a gallon, so the cost to the family could be around $100 per infant.[24] Distillation of water will also eliminate nitrate, as well as bacteria, the suspected contributing factor to methemoglobinemia. Home distillation units are available (see Chapter 17).

11. FLUORIDATION: A Dangerous Gamble

"Smooth runs the water where the brook is deep."
— William Shakespeare

The addition of fluoride to public water supplies, has been the subject of continued debate in both scientific and political arenas since the Public Health Service encouraged communities to adopt the practice in 1950. Today fluoride is added to the water supplies of 57% of the people in the United States.[1]

The fluoridation of public drinking water supplies is a classic exercise in preventive medicine, done for the sole purpose of reducing the widespread problem of tooth decay. Fluoridation is endorsed and supported by most major health organizations including: the American Medical Association; the American Dental Association; the American Institute of Nutrition; and the American Society for Clinical Nutrition. Proponents of fluoridation herald it as a safe and effective means of preventing tooth decay. It was the goal of the Centers for Disease Control that by 1990 almost all people on community water systems would drink fluoridated water.[2]

Optimum levels for fluoride in water are considered to be between 0.7 and 1.2 ppm. According to the Surgeon General, "an optimum level of fluoride in drinking water is best defined as that concentration which provides the highest level of protection against dental cavities consistent with the minimal prevalence of clinically observable dental fluorosis."[3] Of course, the total fluoride ingested is dependent on the total amount of water that one drinks. Those living in warmer sections of the country generally drink more water due to higher average year-round temperatures. The Centers for Disease Control is, however, less concerned about the overfeeding than the underfeeding of fluoride, because underfeeding would bring about a reduction in the dental benefits.[4]

Recent studies have raised doubts about the long-held belief in the dental benefits of fluoridation. Dr. John A. Yiamouyiannis, a noted researcher and opponent of fluoridation, compared school children in both fluoridated and non-fluoridated areas. Using data collected by the National Institute of Dental Research he found that the average decay rates were nearly identical.[5] Another study by the New York State Department of Health compared decay rates in 7–14 year olds in Newburgh, which has been fluoridating its water since 1945, and Kingston, a city that never used fluoridation. Results showed that since 1955, the decay rate dropped more in Kingston than in fluoridated Newburgh.[6] A survey of scientific reports found major reductions in dental cavities in the last 30 years in areas of eight industrialized countries irrespective of whether the area was fluoridated or not.[7]

Health Effects of Fluoride in Drinking Water

Dental fluorosis is the most widely agreed upon negative effect of excess fluoride.[8] The appearance of dental fluorosis can range from hardly noticeable or minor cosmetic changes to the most severe cases which are characterized by dark brown stains and pitting or mottling of the teeth. The degree of fluorosis will depend on the amount of fluoride and length of time it is ingested. Dental fluorosis is labeled a cosmetic effect rather than a health effect so it is not considered in the regulation of fluoride in water supplies.[9]

Another health effect of excess fluoride is skeletal fluorosis. It is characterized by rheumatic attack, pain, and stiffness. The effects of skeletal fluorosis have been seen to a greater extent in other countries where fluoride levels up to 40 ppm have been observed. Only two cases have been reported in the U.S. due to drinking water. In both instances large amounts of water, as well as tea, were consumed.[10] Tea contains more fluoride than most other foods. According to the Surgeon General, skeletel fluorosis effects are experienced from ingestion of 20 mg of fluoride daily for a period of 20 years or more.[11] This corresponds to a 10 ppm drinking water level if you drink two liters of water per day as is considered the average consumption by EPA estimates.

The debate continues in the United States and around the world over other health effects of fluoridation. Dental journals are filled with studies showing how fluoridation is preventing tooth decay in areas where it is practiced. However, there are comparatively few studies on the long-term human health effects of ingestion of artificially fluoridated water.

According to EPA officials, conflicting studies on the health effects of fluoride in drinking water were the basis for their conclusion that inadequate evidence exists to severely restrict fluoride levels in drinking water. The EPA believes the risks to the population from a fluoride level of 4 ppm (the MCL) are negligible.[12] Despite its review of eleven scientific papers concluding that fluoride is oncogenic, the EPA determined, based on conflicting studies and public comments, that there was lack of information to suspect fluoride has the potential to cause cancer in humans.[13] The situation is much the same in the review of the mutagenicity of fluoride.

In 1975, Dr. Dean Burk and Dr. John A. Yiamouyiannis compared the cancer death rates of the ten largest fluoridated U.S. cities to the cancer death rates of the ten largest non-fluoridated U.S. cities for the years 1940 to 1970. Between 1940 and 1952, during the time which no cities were fluoridated, the cancer death rates between the two groups of cities were similar. However, by 1969, after years of fluoridation, data showed the fluoridated cities had an average cancer death rate of 220–225 per 100,000 people. The average cancer death rate for the non-fluoridated cities was 195–200 for every 100,000 people.[14]

Burk and Yiamouyiannis, both noted scientists, have gone against the grain of the general medical and scientific acceptance of fluoridation. According to Burk, one-tenth of all cancer deaths may be linked to fluoridation of drinking water, amounting to about 40,000 cancer deaths per year.[15] Yiamouyiannis estimates that those drinking artificially fluoridated water have between a 5% and 15% increased risk of dying of cancer. This means 10,000 to 20,000 cancer deaths yearly may be due to the fluoridation of drinking water.[16] According to Burk, the latency period for fluoride-induced diseases is four to seven years, compared with the ten to fifteen years or more, for cancers from other sources.[17] In other words, four to seven years after the initiation of fluoridation the effects would be expected to surface. The controversy over the handling of data in the studies of fluoridation and cancer by Burk and Yiamouyiannis led the National Cancer Institute to review the available evidence in the 1970s and conclude that no trends in health effects were observed from the ingestion of water containing fluoride.[18] Just this year, however, the National Toxicological Program, reported that preliminary evidence shows elevated levels of fluoride caused cancer in male rats.[19]

Studies have determined that fluoride could be a factor contributing to other health effects as well. Dr. George L. Walbot reviewed scientific literature on an apparent association between fluoride and Down's syndrome. He determined that there is an elevated incidence of Down's syndrome in newborns in areas with high fluoride levels in drinking water.[20] Scientists from Sri Lanka determined that 1 ppm of fluoride in water (the level associated with fluoridation) leaches almost 220 ppm of aluminum into water when it is boiled for 10 minutes in aluminum cooking pots. This is 1,000 times the aluminum leached by non-fluoridated water. Longer boiling times, according to the researchers, could raise the aluminum level to about 600 ppm.[21] Although a specific cause has not been determined, aluminum has been associated with Alzheimer's Disease.

In his book, ***Fluoride: The Aging Factor,*** Yiamouyiannis documents several health effects of excess fluoride which lead to premature aging:[22]

- weakening of the immune system;
- the breakdown of collagen which comprises 30% of the body's protein and is a constituent of ligaments, tendons, muscles, cartilage, skin, bones, and teeth. Further, the production of imperfect collagen due to excessive fluoride is, according to Yiamouyiannis, a contributor to arteriosclerosis;
- inhibition of DNA repair and synthesis, thereby potentially leading to genetic or chromosome damage; and
- interference with the body's enzyme operation.

Fluoride is believed to inhibit enzymes that repair DNA. Impairment of DNA repair mechanisms lets injured cells reproduce, resulting in chromosonal injury. According to Yiamouyiannis, fluoride may interfere with DNA building block material causing cancer.[23] Other researchers have reported evidence that fluoride could form hydrogen bonds with DNA, disrupting it and thereby leading to mutations or tumors.[24]

A Need for Fluoridation?

Were it not for the effects the popular American diet has on teeth, fluoridation would not be perceived as necessary by so many people. The large quantities of sugar and refined carbohydrates consumed by the American public is the major contributing factor in the prevalence of tooth decay — not the absence of fluoride in drinking water. Changing one's diet would be the logical solution to the problem. Even though drinking water in the United States is some of the most heavily fluoridated, the U.S. continues to have one of the highest rates of tooth decay.[25]

Studies have found dental cavities in primitive peoples almost non-existent due to their natural diet.[26] Modernization of these people and conversion to the traditional American diet brought about a significant incidence of tooth decay. A study of isolated cultures found an average of one dental cavity per 100 teeth checked. However, after only a short period of contact with the culture and diet of the Western world, the number increased to 30 cavities per 100 teeth.[27]

Society has come to expect a quick fix or easy solution to the problems we create for ourselves. When it comes to tooth decay, fluoridation was made to order. However, this is a deliberate addition of a foreign substance — a pollutant — to water explicitly for ingestion by humans. Fluoride is a waste product generated in a number of different industrial processes, including the manufacture of aluminum, glass, bricks, steel, and phosphate.

Because the EPA can find no justification within the scientific literature to regulate a constituent like fluoride does not mean that their interpretation is conclusive. Municipalities and the public often accept the opinions of federal agencies and the medical community at face value. Even fence-sitters in the controversy should agree that with so many unanswered questions, it may not be prudent to add this substance to our water supplies. Laypeople, including government regulators, often mistakenly assume that the relationship between pollutants and health effects is black and white and well understood. They proceed on a risky course with the meager assurance that a number of interpretations of current data does not indicate health effects. Future evidence may sharply reverse current policy. It would certainly not be the first time a chemical was banned or severely restricted after experience and developments found it hazardous.

Keep in mind that fluoride is known to be acutely toxic at only slightly higher levels. Airborne fluoride was a factor in the deaths of 20 people in Donora, Pennsylvania in 1945,[28] and caused 60 deaths and several thousand illnesses in the Meuse Valley, Belgium in 1930.[29] The effects of low levels of fluoride over a period of time are difficult to prove scientifically. A very fine line exists between hazardous levels and levels considered beneficial to teeth. Dr. William Lappenbusch, noted drinking water toxicologist and former chief of the health effects branch of the EPA's Office of Drinking Water, while supporting fluoridation cautions that the window between the beneficial and detrimental health effects is very small.[30] It is a double-edged sword.

Many scientists do not believe it is necessary to fluoridate drinking water supplies because other sources of fluoride are more than adequate if there is a concern about tooth decay. Toothpaste, mouthwash, and food are all sources of fluoride. Also, fluoride treatments can be obtained for those desiring it. Even the EPA's Union of Professional Workers and Scientists has urged the agency to reevaluate its position on fluoridation.[31]

Simply as a matter of comparison, most countries, particularly in Europe, have either never initiated fluoridation or have stopped the practice due to its potential long term dangers and environmental effects.[32] Roughly only 2% of Europe's population drinks fluoridated water.[33] Los Angeles, Honolulu, San Diego, San Antonio, and other cities have voted down referendums for fluoridation. The decision to fluoridate can be made by health officials or city councils and not by referendum in some areas of the country. However, since 1950, 60% of the 2,000 referendums on the fluoridation of water in the U.S. have been voted down.[34]

It is, of course, possible that history may prove that the fluoridation of drinking water is one of the greatest preventive measures of all time. In the face of current evidence, however, it is unlikely. Is it worth the risk of biding our time and waiting to find out?

Testing for Fluoride

There are two methods for testing fluoride in water — the **SPADNS method** (the traditional method) or the *specific ion electrode method.* Both are sensitive enough to detect fluoride at low levels. The SPADNS method has a number of interferences that could result in a false determination. Low levels of other ions like chlorine, phosphates, iron, and aluminum could interfere with the test. With the specific ion electrode method, the possibility of interference is much lower and it is therefore considered more accurate.[35] The use of the specific ion electrode method could also be an indicator of a more sophisticated lab. A laboratory offering this test may be more concerned about the quality of their work. All that is required is that the electrode be inserted in the sample and a digital reading is given by the instrument. The specific ion electrode method measures in the 0.1 to 10 ppm range. The SPADNS method measures in the 0.1 to 2.0 ppm range. To measure higher levels would require dilution of the sample, and this could lead to errors.

Test Results & Maximum Contaminant Level (MCL)

In 1987, the EPA revised the interim Maximum Contaminant Level

(MCL) for fluoride to 4.0 ppm to protect against rare, crippling, skeletal fluorosis.[36] They also promulgated a secondary standard to protect against dental fluorosis. The secondary standard is not enforceable, but water systems that have fluoride over 2.0 mg/L are required to inform their customers. According to the EPA, this is merely a guide directed primarily to systems that have naturally occurring high fluoride levels. Fluoride levels in excess of the Maximum Contaminant Level are usually due to natural occurrence related to the geology of the area.[37] South Carolina, for example, has many areas like this. Generally, the deeper the ground-water the higher the fluoride concentration.

Reducing Fluoride in Drinking Water

Reverse osmosis (RO) is one of the best methods, both economically and technologically, for fluoride removal. RO removes 90% to 95% of the fluoride present and is capable of treating many other contaminants as well. RO is available in point-of-use (faucet) and point-of-entry (whole house) home treatment devices. Distillation also removes fluoride quite effectively (see Chapter 17).

12. SYNTHETIC ORGANIC CHEMICALS

"I believe firmly that we cannot afford to give chemicals the same constitutional rights that we enjoy under the law. Chemicals are not innocent until proven guilty."

— Russell Peterson, Former
Chairman, U.S. Council
on Environmental Quality

Most of the previous chapters have focused on **inorganic chemicals**, metals and other non-metals that do not contain carbon. Analyzing for inorganics is relatively inexpensive. They typically, but not always, exhibit acute health effects and their source is often easily determined. **Organic chemicals** are nearly the opposite. They contain, among other things, varying amounts of carbon.

Organic chemicals can be divided into two different and distinct categories, natural and synthetic. Natural organics occur in water as result of the normal decay of plant life characteristic of forests and grasslands. These compounds are rather complex and have not been widely studied. However, plants have been dying and decaying for millions of years and the resulting compounds are an integral part of aquatic ecosystems and are not considered harmful.

Synthetic organic chemicals (SOCs), on the other hand, are not found naturally in the environment. When man realized he could create new chemicals in the laboratory by introducing constituents that would not ordinarily meet under natural conditions, the chemical age was born. Synthetic organic chemicals are foreign to natural systems. Since they are artificially created, biological systems do not have the ability to breakdown or degrade these invading chemicals. Detection of SOCs is generally more expensive and difficult than inorganics testing, and exposure to them usually results in chronic health effects such as cancer. SOCs are also generally toxic at much lower levels than inorganic chemicals and metals.

The synthetic organic chemical industry rose from obscurity in the early part of this century to the powerful economic and political force it is today. Synthetic chemical production began to spiral upwards as a movement from coal tar-based chemicals to petroleum-based chemicals took place. In 1930, the SOC industry was in its infancy and production was a mere few hundred thousand pounds yearly.[1] Today, in the aftermath of the post-World War II boom in technological and economic development, production is over 387 billion pounds per year.[2] Most of the more than four million chemicals registered by the American Chemical Society and the Chemical Abstracts Service are SOCs.[3] About 60,000 of these are in commercial production.

SOCs can be further divided into two categories, **volatile** and **nonvolatile.** Volatile organic chemicals (VOCs) are small, lightweight compounds that can easily dissipate from water into the air. Examples of VOCs are THMs, trichloroethylene (TCE), and gasoline products like benzene. Nonvolatile organic chemicals (NVOC) are usually heavier and, like metals, will drop to the bottom of waterways. NVOC levels are usually highest in the sediment of waterways. Examples of NVOCs are PCBs, DDT, and phthalates. Some SOCs are **halogenated**. This means they contain chlorine or bromine, elements that are called **halogens.** Halogenated SOCs are some of the most significant pollutants.

The majority of SOCs that have been tested have exhibited some sort of negative effect. However, most have not been tested at all. According to a report by the National Research Council, very little or no toxicity data exists for 64% of pesticides, 74% of cosmetic ingredients, and almost 79% of all chemicals in use in the United States.[4] This significant lack of toxicity data makes it inherently difficult for government regulators to make hard decisions about safe levels of chemicals. Often the decision not to regulate a chemical in drinking water is based on the fact that no information is available on its toxicity. Environmental personnel in the field also find it difficult to make good quality judgments about the safety of a water supply. When a homeowner with a contaminated well wants to know if his water is safe to drink, an accurate answer may not be available. Not everything causes cancer, but synthetic chemicals have the potential for a wide variety of serious health effects.

The health effects of SOCs vary depending on the specific chemical and are as broad as the category itself. They include carcinogenic, mutagenic, teratogenic, and oncogenic effects. The extent of a particular SOC's hazard to human health depends on two things, how toxic it is and the magnitude of human exposure. Highly toxic chemicals will not be harmful if contained, and a chemical with a lower toxicity can be a hazard if exposure is heavy. Media news of hazardous waste or toxic chemicals most often involves SOCs.

Further adding to the uncertainty over safe levels of drinking water is the contribution of skin absorption and inhalation to total SOC exposure. Hot showers, where SOCs easily volatilize and the confined area can result in the breathing of high levels, are of the greatest concern with respect to inhalation. One study determined that skin absorption accounted for 29–91 percent of the total level of exposure.[5] The researchers concluded that skin absorption has been underestimated as a pathway of exposure.

Regulation of SOCs in Drinking Water

The EPA lists about 129 contaminants as *priority pollutants* (Table 12-1). SOCs account for 114 of them. Priority pollutants are generally the chemicals most commonly used by industry, and therefore most often disposed of as waste. Consequently, they are of most concern in waterways. Under the federal regulations of the Clean Water Act, corporations are required to monitor their effluent discharge to waterways. Typically this testing will involve only priority pollutants. However, many pesticides and other SOCs are not included in the priority pollutant list. Therefore, while many of the common pollutants are covered, it is by no means a complete list. A study of 4,000 wastewater samples from industry and publicly-owned sewage treatment plants found that of the 50 most frequently found compounds only 14 were priority pollutants.[6]

Compared to the number of priority pollutants, not to mention the huge amount of chemicals in production, the number of chemicals regulated in drinking water has traditionally been meager. A few metals, radionuclides, and six pesticides were the only contaminants regulated in drinking water for many years. In 1979, a Maximum Contaminant Level (MCL) for Trihalomethanes (THMs) was set. More recently, MCLs were

Table 12-1
The Priority Pollutants

Volatile Organic Compounds

acrolein	2-chloroethylvinyle ether	bromoform
acrylonitrile	chloroform	dichlorobromomethane
benzene	1,1-dichloroethylene	trichlorofluoromethane
carbontetrachloride	1,2-trans-dichloroethylene	dichlorobromomethane
chlorobenzene	1,2-dichloropropane	chlorodibromomethane
1,1-dichloroethane	1,3-dichloropropene	tetrechloroethylene
1,2-dichloroethane	ethylbenzene	toluene
1,1,1-trichloroethane	methylene chloride	trichloroethylene
1,1,2-trichloroethane	methyl chloride	vinyl chloride
1,1,2-2-tetrachloroethane	methyl bromide	bis (chloromethyl) ether
chloroethane		

Base-Neutral Extractable Organic Compounds (EPA 625)

acenaphthene	4-bromophenyl phenyl ether	dimethyl phthalate
benzidine	bis (2-chloroisopropyl) ether	benzo(a)anthracene
1,2,4-trichlorobenzene	bis (2-chloroethoxy) methane	benzo(a)pyrene
hexachlorobenzene	hexachlorobutadiene	3,4-benzofluoranthene
hexachloroethane	hexachlorocyclopentadiene	benzo(k)fluoranthene
bis (2-chloroethyl) ether	isophorone	chrysene
2-chloronaphthalene	naphthalene	acenaphthylene
1,2-dichlorobenzene	nitrobenzene	anthracene
1,3-dichlorobenzene	N-nitrosodimethylamine	benzo(ghi)perylene
1,4-dichlorobenzene	N-nitrosodiphenylamine	fluorene
3,3-dichlorobenzidine	N-nitrosodi-n-propylamine	phenanthrene
2,4-dinitrotoluene	butyl benzyl phthalate	dibenzo(a,h)anthracene
2,6-dinitrotoluene	di-n-butyl phthalate	ideno(1,2,3-cd)pyrene
1,2-diphenylhydrazine	di-n-octyl phthalate	pyrene
fluoranthene	diethyl phthalate	bis (2-ethylhexyl)phthalate
4-chlorophenyl phenyl ether		

Acid Extractable Organic Compounds (EPA 625)

2,4,6-trichlorophenol	pentachlorophenol	4,6-dinitro-o-cresol
parachlorometa cresol	2,4-dimethyphenol	2,4,-dichlorophenol
2-chlorophenol	4-nitrophenol	phenol
2-nitrophenol	2,4-dinitrophenol	

Pesticides and PCBs

adlrin	endrin	PCB-1254
dieldrin	endrin aldehyde	PCB-1221
chlordane	heptachlor	PCB-1248
4,4-DDT	heptachlor epoxide	PCB-1248
4,4-DDE	α-BHC	PCB-1260
4,4-DDD	β-BHC	PCB-1016
a-endosulfan	δ-BHC	toxaphene
b-endosulfan	γ-BHC	2,3,7,8-tetrachlorodibenzo-p
endosulfan sulfate	PCB-1242	diozin (TCDD)

Metals

antimony	mercury
arsenic	nickel
beryllium	selenium
cadmium	silver
chromium	thallium
copper	zinc
lead	

Miscellaneous

asbestos
phenols
total cyanides

Source: U.S. EPA

set for eight volatile SOCs, benzene, carbon tetrachloride, 1,2,-dichloro-ethane, trichloroethylene, p-dichlorobenzene, 1,1,-dichloroethylene, 1,1,1,trichloroethane, and vinyl chloride. Under the Safe Drinking Water Act (SDWA) amendments of 1986, a number of organics, including several pesticides, were scheduled for regulation (Table 12-2). MCLs will be set for these in the near future. In addition, all water supplies will be required to monitor for additional SOCs (Appendix II). Monitoring will depend on the water supply. Surface water systems must sample at least four times per year, groundwater supplies at a minimum of every five years. Keep in mind that this does not apply to very small water systems including private wells.

TRIHALOMETHANES & DISINFECTION BYPRODUCTS

The practice of adding chlorine to drinking water to remove viruses, bacteria, and other waterborne disease-causing organisms has been utilized in the United States since 1908. The Boonton Waterworks serving Jersey City, New Jersey, was the first system in the United States to use chlorine to disinfect their drinking water. The decision to chlorinate was challenged in court, but the judge reviewing the case concluded that the disinfection solution was effective and left no harmful substances in the water.[7] By 1918, the practice was in use in most large cities in the United States. The typhoid death rate, already on the decline due to filtration of public water supplies, was further reduced by widespread chlorination.

In 1974 it was discovered that volatile organic chemicals called Trihalomethanes (THMs) are formed in the water. They are the products of the reaction of chlorine with the natural organics from decayed plant life occurring naturally in the water.[8] THMs is a collective term for the four volatile compounds which are created: chloroform, bromodichloro-methane, bromoform, and chlorodibromomethane. In 1976, the National Cancer Institute (NCI) released a study showing chloroform was car-cinogenic to rats and mice.[9] Although chloroform is not the only THM, the other three are not *usually* found at concentrations as high as chloroform. While chlorine is the disinfectant with the most widespread use, bromine is also a disinfectant which can be utilized. This would, however, simply shift the THMs to increase the brominated organic THMs, probably without changing the total THM result.

An MCL of 100 ppb for all four (total) THMs was promulgated in 1979.[10] The selection of this level is based largely on the public health benefits of chlorination, which provides biologically safe water. According to Joseph Cotruvo, director of Criteria and Standards Division of EPA's

Table 12-2
SOC Contaminants Required to be Regulated
under the SDWA of 1986

Volatile Organic Chemicals

Trichloroethylene	Benzene
Tetrachloroethylene	Chlorobenzene
Carbon tetrachloride	Dichlorobenzene
1,1,1-Trichloroethane	Trichlorobenzene
1,2-Dichloroethane	1,1-Dichloroethylene
Vinyl chloride	trans-1,2-Dichloroethylene
Methylene chloride	cis-1,2-Dichloroethylene

Other Organics

Endrin	1,1,2-Trichloroethane
Lindane	Vydate
Methoxychlor	Simazine
Toxaphene	PAHs
2,4-D	PCBs
2,4,5-TP	Atrazine
Aldicarb	Phthalates
Chlordane	Acrylamide
Dalapon	Dibromochloropropane (DBCP)
Diquat	1,2-Dichloropropane
Endothall	Pentachlorophenol
Glyphosate	Pichloram
Carbofuran	Dinoseb
Alachlor	Ethylene dibromide (EDB)
Epichlorohydrin	Dibromomenthane
Toluene	Xylene
Adipates	Hexachlorocyclopentadiene
2,3,7,8-TCDD (Dioxin)	

*Source: U.S. EPA, **54 Federal Register**, 22 May 1989, p. 22141.*

Office of Drinking Water, the MCL for THMs should not be interpreted as an absolutely safe level, but the level attainable with the water treatment technology generally available since 1974.[11] The National Academy of Sciences also believes the 100 ppb standard should be reduced.[12] The EPA will likely make standards for THMs and other byproducts of disinfection more stringent sometime in the future.[13]

A survey of large municipal water companies determined that total THMs averaged 42 ppb.[14] The survey also determined that 60% of the utilities would fail to meet the total THM standard if it were reduced to 25 ppb. According to federal regulations, systems with under 10,000 people are required to meet the THM level at the discretion of the state government.[15]

Unfortunately, THMs are not the end of the chlorination story. In addition to the volatile THMs, heavier nonvolatile chlorinated organics are also formed. It is likely that there are more of these nonvolatile chemical compounds created because most of the natural organic material in drinking water is nonvolatile. An estimated 90% to 95% of the organic material present in treated and untreated water is nonvolatile.[16] Consequently, it is believed that the chlorinated nonvolatile chemicals created during disinfection similarly outweigh the level of THMs. It is more difficult to specifically identify these heavier chlorinated organics than it is to identify volatile organics. Therefore, many are only tentatively identified.[17]

Eighty percent of the smaller municipal water systems utilize groundwater that is low in natural organics.[18] Groundwater, as a general rule, does not contain the decayed natural organic material at the level that surface water does. Surface water encounters plant, animal, and forest organic products in its travels to the drinking water treatment plant. Groundwater does not contact these surface products. Consequently, it is less prone to high levels of chlorinated organics as a result of disinfection. Also, in many instances, groundwater is not chlorinated. It generally does not need to be because, unlike surface water, it is not subjected to heavy sewage discharge.

Health Effects of THMs and Disinfection By-Products

Several studies have found a statistically significant relationship between the presence of THMs in drinking water and cancer.[19] These studies support an association between bladder, colon, and rectal cancer and the ingestion of water containing THMs.[20] However, as discussed previously, THMs are only a very small part of total organic contaminants present in drinking water systems. Any conclusions in terms of health effects of disinfection byproducts must consider both volatile and nonvolatile constituents.

Other studies have determined significantly increased risks of rectal or urinary and gastrointestinal cancers associated with not only THMs, but chlorinated water in general.[21] Data analyzed by the National Cancer Institute determined a risk of bladder cancer associated with the level and frequency of tap water ingestion. This was particularly true among long-term residents of areas using chlorinated surface water.[22]

One study found cancer mortality rates for stomach, bladder, and other cancers were greater for white males in counties served by surface water supplies.[23] The researchers found these differences in mortality rates were independent of other confounding factors like urbanization and industrial activity. The study indicates that organic chemical contaminants in surface water resulting from chlorination or industrial pollution may be related to the higher cancer mortality.

Several studies have also pointed to chlorination as contributing to the mutagenic activity of water by the formation of nonvolatile halogenated chemicals.[24] Measurements of extracts of these unknown chemicals from disinfected drinking water samples have been positive in mutagenicity tests. Some compounds with mutagenic or carginogenic properties have been identified in drinking water, but researchers have been unable to link the mutagenic activity in water samples to specific chemicals. Therefore, scientists are not certain what contribution the specific chemicals identified to date make to the overall mutagenic activity in drinking water.[25]

Despite the lack of a clear and conclusive link between byproducts of disinfection and cancer or mutagenic effects, scientists generally agree that continued investigation is necessary in view of the potential significance of the findings to date. As the National Academy of Sciences points out, considering the frequency of chlorinated drinking water consumption, only a small increased risk could mean a considerable level of disease.[26] Many agree that because little is known about the extent and effects of contamination, attempts must be made to reduce these toxic organic chemicals in drinking water. But the microbiological quality must also be maintained. The problem is how to meet both requirements simultaneously.

Testing for THMs

For those concerned about THMs, it may be more advantageous to request a complete *volatiles scan* (see Testing for Volatiles) rather than testing for only four chemicals. Those on a community system can request lab reports on THMs from the water company. This will also reveal the extent of halogenated nonvolatile chemical formation. Keep in mind, however, that THMs and heavier chlorinated organics can form in the distribution system after the point where the water company tests for them. The level of THMs will likely be higher at the tap than at the treatment plant. So even if one were a firm believer in the safety of Maximum Contaminant Levels (MCLs), the THM concentration may be in compliance at the plant but not coming out of the faucet. THM levels will generally be the highest in those sections of the distribution system where water spends the longest time. [27]

VOLATILE ORGANICS

Only about ten percent of all possible organic water contaminants are volatile. However, ninety percent of the volatile organic chemicals (VOCs) that exist in drinking water have been identified. [28] From a testing standpoint, that is good news. It means that in the volatile area, scientists know a great deal about specific chemical pollutants and the extent of contamination. In addition, a large percentage of commonly used industrial solvents with a high potential for water contamination are volatile chemicals.

Volatile SOCs have generally been the most prevalent chemicals found in groundwater. Twenty-nine VOCs were surveyed by the EPA Office of Drinking Water in the 1984 Groundwater Supply Survey. At least one VOC was found in 22% of the water supplies serving less than 10,000 people, and in 37% of water supplies serving more than 10,000 people. [29] The three most frequently found compounds were trichloroethylene, tetrachloroethylene, and 1,1,1 trichloroethane.

Sampling for Volatiles

Taking a sample for VOC analysis is particularly critical. A glass bottle with a teflon-lined cap is required. This can be obtained from the laboratory that will do the testing. The way the bottle is filled is important. Do not fill it too fast; make sure the water does not bubble and splash. Fill the bottle to overflowing, so that the water beads up over the top. Then place the cap carefully over the top so that *no air bubbles* get into the bottle. The inside of the bottle must be all water. Turn the bottle upside down to check for air bubbles. If any bubbles are present, start over.

If there are air bubbles in the sample bottle, the volatile organics can transfer from the water into the air inside the bottle in the time before the sample is analyzed. When the sample bottle is opened in the lab, the contaminants are "gone with the wind." Analyzing a volatiles sample with a bubble in it is useless, a waste of time and money. Results are invalid, erroneous, and can give a false sense of security about water quality.

No chemical preservative is required, but keep the sample refrigerated before it is analyzed. It should be put on ice if it is to be shipped. The sample should be analyzed within 14 days. Always date the sample so the lab is aware of how long the sample has been held.

Testing for VOCs and Other SOCs

The similarity of certain chemicals determines how they are tested. Often, similar chemicals are easily analyzed by the same method. For example, all the common chemicals that are volatile are anlayzed by the same methods. Certain classes of pesticides are also analyzed together because they have similar characteristics. Consequently, it is often just as easy for a lab to look for all the chemicals in a particular method when testing a sample as it is to look for only a single chemical.

As mentioned earlier, volatile chemicals are those that evaporate easily. That is, they can move or dissociate from water into the air over time. In the laboratory this can be speeded up for analytical purposes by *purging* the water sample. The water sample is put in a chamber and nitrogen is bubbled through. Volatile organic chemicals in water samples are released into the bubbles. These bubbles of gas containing the volatile chemicals rise out of the water and are collected on a *trap.* This trap is then analyzed to determine exactly what specific chemicals are present. This method, appropriately called *purge and trap,* is used in identifying specific volatile pollutants. The lab may use other terms to describe this method. They are *volatiles scan, volatiles organics analysis* (VOA), and *volatiles.* The lab might use any of these to describe the method for testing volatiles.

Probably the best way to communicate a request for VOC analysis is by using the EPA method numbers. Series 600 volatile methods, 601 (halogenated) and 602 (nonhalogenated), have been generally available at commercial labs for quite some time. Their primary purpose is the analysis of industrial wastewater. However, their applicataion to drinking water is acceptable, and these methods have been utilized extensively to test potable water. Recently, 500 series methods, 502 for halogenated and 503 for nonhalogenated volatiles, have been developed. They cover many of the same chemicals but are used specifically for drinking water testing. Another difference between the 500 and 600 series is the sample volume. Larger sample volumes are used in the 500 series to increase the sensitivity and achieve a lower *detection limit,* the lowest level of a contaminant that can be detected. The performance criteria built into the 500 series methods are more stringent, and the EPA does not accept 600 methods for compliance by water utilities with monitoring regulations.[30] However, any lab with good quality control practices will give valid results regardless of what series they use.

The majority of analytical methods for synthetic organic chemicals use a Gas Chromatograph (GC) to separate the different chemicals in a

Figure 12-1
Organic Pollutants Are Extracted From Water Using a Solvent

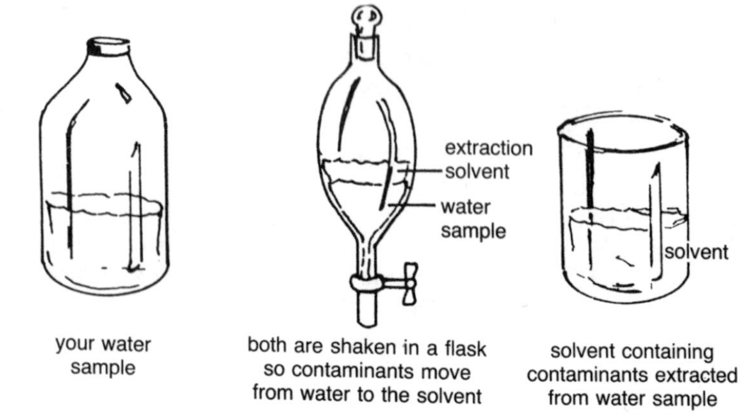

extraction solvent

water sample

solvent

your water sample

both are shaken in a flask so contaminants move from water to the solvent

solvent containing contaminants extracted from water sample

The Gas Chromatograph

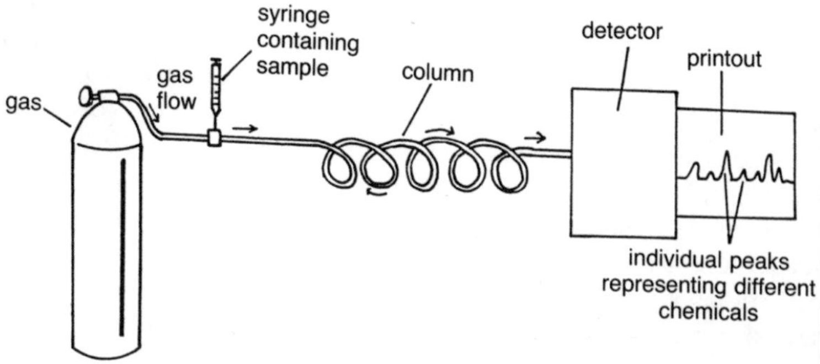

syringe containing sample

detector

printout

gas flow

column

gas

individual peaks representing different chemicals

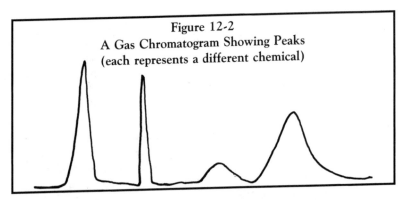

Figure 12-2
A Gas Chromatogram Showing Peaks
(each represents a different chemical)

sample (Figure 12-1). First, the contaminants are extracted from the water sample using a solvent. The solvent and water sample are mixed in a funnel to allow the contaminants in the water to move into the solvent. The solvent is then injected into the GC where a gas forces the sample through a column which separates all the contaminants.

How long a particular chemical is retained on the column (some move slower than others) will give a clue to exactly what it is. Thus, each chemical in the method has a ***retention time*** — how long before it comes out of the column. A ***detector*** at the end of the column measures the amount of each component in the sample and prints out a ***peak*** if a contaminant is present (Figure 12-2). A match-up between the retention time of a sample peak and the retention time of a known chemical will indicate the chemical present in the sample. The presence of any sizable peak is usually a sign of contamination. The size of the peak is an indicator of how much of the contaminant is present in the sample. The larger the peak, generally, the higher the concentration of the contaminant. Companies required by law to have their wastewater effluents monitored typically find many peaks on their chromatogram. However, when testing drinking water, particularly well water, no peaks should be found.

In the analysis of drinking water, the volatiles scan can be used as a screening procedure to search for *any* volatile contaminants. To achieve as comprehensive an assessment of water quality as possible, one would want to look for *any* chemicals, not just a particular one. It will probably cost about the same to look for one chemical as it will to run the entire scan.

If a scan is requested, be sure the lab points out any and all peaks, regardless of whether they can identify the peak as being a specific compound or not. Always request a copy of the GC printout. While a method is designed to identify specific chemicals, it may pick up other chemicals as well. These will also appear in the form of a peak.

Appendix III lists the specific volatile chemicals covered in the EPA methods. Many labs, however, can go beyond this and quantify additional volatile compounds that may turn up in the analysis. The lab may have the capability to determine the presence and amount of 90 or more volatiles. The EPA 601/602 procedures call for only 36. The lab may ignore an unidentified peak if a comprehensive screen is not requested. However, do not search far and wide for a laboratory with a large data base. All labs following the 601/602 method will be able to pick up peaks. All you need to know initially is whether or not the water is contaminated. If it is, you can go back to the lab to find out what the specific pollutant is.

OTHER NONVOLATILE SOCs and TESTING METHODS

While many volatiles found in drinking water have been identified, the nonvolatiles are another matter entirely. Even though they account for about 90% of the oganics in drinking water, only 5 to 10% have been identified.[31] In addition to volatiles testing methods, there are a number of nonvolatile SOC tests which are widely available (Appendix III). If you still feel there is a problem after no contamination has been detected in tests for volatiles, TOX (see chapter 15), and metals, these nonvolatile tests might be considered. They are also helpful when contamination is found in these initial tests and you want to know what else may be present.

Nonvolatile organic chemicals are in widespread use and therefore are believed to be found throughout the environment. However, they have not generally received the attention by scientists and regulators that volatiles have. The most common nonvolatiles are PAHs, phthalates, and phenols. All can be tested using conventional GC methods, EPA 610 (or 550), EPA 606 (or 506), and EPA 604 respectively.

Polyaromatic Hydrocarbons (PAHs)

Polyaromatic hydrocarbons (PAHs) are a class of compounds formed mainly during the combustion of organic materials like wood, coal, fuels, and other chemicals. Table 12-3 lists common PAHs. Some are known or suspected carcinogens. In 1775 in England, it was first discovered that chimney sweeps had higher rates of scrotal cancer than the rest of the population. The source of the cancer was traced to the chimney soot. It was not until 1933 that benzo(a)pyrene, a PAH, was isolated from coal dust and found to metabolize in the body to form more than one carcinogen.[32]

Other than atmospheric deposition of airborne particles containing PAHs, they can enter water sources from:

- road surface and tire wear which contributes to PAH runoff after precipitation;
- vehicle emmissions; and
- industries involved in oil refining, plastics production, and dyestuffs manufacture that use petroleum products or utilize furnaces.[33]

PAHs are likely to be more prevalent in surface water than groundwater. However, anthracene and phenanthrene have been reported in wells in New York state.[34] Consequently, both surface water and groundwater users should be aware of them.

Table 12-3
Common PAHs

Acenaphthene	Chrysene
Acenaphthylene	Dibenzo(a,h)anthracene
Anthracene	Fluoranthene
Benzo(a)anthracene	Fluorene
Benzo(a)pyrene	Indeno(1,2,3-cd)pyrene
Benzo(b)fluoranthene	Naphthalene
Benzo(ghi)perylene	Phenanthrene
Benzo(k)fluoranthene	Pyrene

Phthalates

Plasticizers, including many phthalates, are possibly the most widespread of all environmental pollutants.[35] Phthalates (Table 12-4) are a group of chemicals used in plastics to make them soft, flexible, and workable, qualities that make plastics attractive and marketable products. One of the chief products using phthalates as plasticizers is PVC (polyvinyl chloride), which contains as much as 40% phthalates.[36] In a wide variety of other products (Table 12-5), phthalates may comprise up to 60% of the total weight of the plastic.[37]

The heavy usage and soaring disposal rates of plastics could be causing widespread pollution and environmental damage as phthalates slowly leach or migrate from the plastics over time.[38] Phthalates have been detected in ground and surface drinking water sources,[39] and have been found in the sediments of waterways.[40] The phthlate level in waterway sediments has been found to coincide with phthlate production. As plastic use and

---

Table 12-4
Common Phthalates

Bis (2 ethylhexyl) phthalate	Dimethyl phthalate
Butyl benzyl phthalate	Diethyl phthalate
Di-n-butyl phthalate	Di-n-octyl phthalate

---

disposal has increased, so has the accumulation of phthlates in these sediments. Sources include: sewage treatment plants; paper and textile mills incinerators; and plasticizer plants.[41] The EPA has reported phthalates in surface waters in the parts-per-billion range, with even higher levels near industrial areas.[42] Of the ten most commonly found contaminants in a survey of 112 organic chemicals in groundwater supplies in New York State, four were phthalates.[43]

Health effects have been associated with a number of phthalates. Dimethyl phthalate (widely used in insect repellents) has been implicated in increasing the risk of skin cancer and mutagenesis.[45] Other phthalates are linked to liver damage.[46] Bis (2-ethylhexyl) phthalate is labeled carcinogenic by the National Cancer Institute.[47]

Table 12-5
Uses of Phthalates[44]

Construction Materials — cable and wire coverings, flooring, weatherstripping, and pool liners.

Furnishings — upholstery, wallcoverings, housewares, appliances, garden hose

Automobiles — seat covers and interior upholstery, car mats, and car tops

Garments — shoes, outer apparel, and baby products

Food and Medical — food wrap, medical tubing, and IV bags

Other Uses — cosmetics, perfumes, insect repellents, pesticide carriers, & industrial oils.

Phenols

Phenols, sometimes called phenolics, are most often used as intermediaries to synthesize a large number of other chemicals. Phenol itself is a widely used chemcial in a variety of industrial and manufacturing processes. The most common phenols are listed in Table 12-6. They can be divided generally into two subgroups: chlorophenols, those containing chlorine; and nitrophenols, those containing nitrogen. Chlorophenols are produced as biocides, insecticides, and wood preservatives and are used as chemical reactants in the manufacturing of dyes, pigments, and resins.[48] Effluent wastes containing phenol are generated by coke ovens, oil refineries, chemical plants, and blast furnaces.[49] Other sources of phenols are listed in Table 12-7. Chlorophenols are also formed when water containing phenol is chlorinated. Also, certain chlorophenols may be formed by the breakdown of herbicides and pesticides.[50]

Table 12-6
Common Phenols

4-chloro-3-methyl phenol	2-Nitrophenol
2-chlorophenol	4-Nitrophenol
2,4-Dichlorophenol	Pentachlorophenol (PCP)
2,4-Dimethylphenol	Phenol
2,4-Dinitrophenol	2,4,6-Trichlorophenol
2-Methyl-4,6-dinitrophenol	

The greatest threat to water contamination by pentachlorophenol (PCP) comes from manufacturing and wood-preservative companies. Up to 18 ppm of PCP was detected in a small stream near a wood preservation facility. The oil floating on the surface of the stream contained 5,800 ppm of PCP.[51] It also can be released to water sources by direct contact of preserved wood with soil or water, but contamination would likely be less than from point sources associated with industrial sites.

In addition to the conventional GC method for testing phenols (604), another method, **total phenols,** exists. Total phenols is a **wet method** which means the chemicals are not extracted from the sample. This method is used to analyze for additional phenols not found in EPA 604.

Table 12-7
Sources of Phenols[52]

• chemical plants	• fabric & fiber manufacturing
• fiberglass manufacturing	• coatings, resins, & adhesives
• plastics industry	• bonding compounds
• housing & auto-products industries	• plywood manufacture & wood
• molding & molding compounds	preserving

GC/MS

Different GC detectors are used in different methods because each is sensitive to different classes of chemicals. The most advanced state-of-the-art detector is the **Mass Spectrometer (MS).** It is coupled with a GC to give one of the most advanced analytical systems available today, **GC/ MS.** What MS does is go one step further than an ordinary GC detector. It takes all individual chemical peaks that the ordinary GC detector finds and fragments them into even smaller parts (Figure 12-3). The **mass spectrum** (GC/MS printout) is a "fingerprint" of the chemical. It is characteristic and unique to each chemical and is more difficult to misidentify, whereas the peak on a conventional GC could possibly be misidentified. This does not mean that MS is "the detector" for all occasions. Many others are used depending on the test method. Nitrogen-Phosphorous detectors, for example, are necessary for pesticides containing nitrogen and phosphorous.

When comparing the MS detector with other "conventional" detectors there are several points to consider. Separation of individual peaks by conventional detectors is not always complete when a number of contaminants or interfering compounds are present in the sample. This can make it difficult to identify and measure the amount of a contaminant. However, in the case of drinking water, there should not be a large number of contaminants. If contaminated, water will likely have more than one contaminant, but pollution to the point of making identification and quantification difficult is usually reserved for wastewaters. Also, when compounds can be easily identified, conventional detectors usually make better measurements than GC/MS. Conventional GC analysis also costs much less. A volatiles scan on GC/MS can cost anywhere from $150–$300, while a similar scan on conventional GC would cost $50 and up. The greater identification capability of GC/MS may be preferable, however, if litigation is a possibility.

GC/MS analysis of water is done through two specific EPA methods, EPA 624 and EPA 625 (see Table 12-1). EPA 625 covers phenols, PAHs, phthalates, some pesticides, and other chemicals. EPA 624 covers volatiles. Some labs may have a new method, EPA 524, which is the same as 624 except that the detection limits for the chemicals are lower. EPA 524 is specifically designed for drinking water.

Another advantage of GC/MS, is that most systems are tied to a computerized data library. This computer can find many contaminants not specifically in the method. A tentative identification of compounds can be made by searching the computer library to find a match of possible

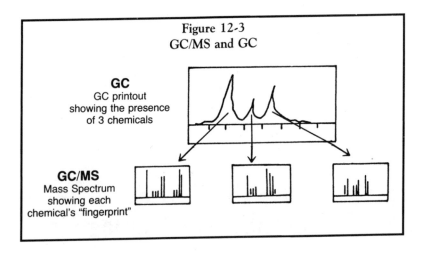

Figure 12-3
GC/MS and GC

GC
GC printout
showing the presence
of 3 chemicals

GC/MS
Mass Spectrum
showing each
chemical's "fingerprint"

contaminants. These are called ***Tentatively Identified Compounds*** (TIC). If a contaminant is found, the computer can tell what it is, and *estimate* how much is there. Because a TIC is not part of the method, its measurement will not be accurate. But, it is still a valuable indication of contamination. A TIC library search is similar to looking for *any peaks* on a conventional GC printout.

Be aware that whether analysis is done by conventional GC or GC/MS, problems with misidentification and quantitative measurement are possible. One study found that, *excluding priority pollutants,* incorrect identification of compounds and inaccuracy in measuring contaminant levels is widespread.[53] The reasons cited were poor data review practices, heavy reliance on automated systems, and time constraints on the chemical analyst.

Reducing SOCs in Drinking Water

Given the number of SOCs, each with unique properties, it is difficult to discuss their treatment in anything but general terms. Refer to Chapter 17 for home treatment methods for the different classes of SOCs.

13. PESTICIDES: Deliberate Damage

"As the review of these chemicals progresses, some chemicals that were once thought to be safe will be found to be unacceptable by modern standards. . . . We should not lose sight of the fact that even new pesticides, which have been carefully scrutinized by modern science and technology and found to be safe, may at some future date be found unacceptable in light of future scientific advances."

— John A. Moore,
Assistant Administrator
for EPA, 1984

Pesticides are chemicals used to control or irradicate unwanted plants and animals. These include: *Herbicides,* which kill unwanted plants and weeds; *insecticides,* used to kill insect pests; *fungicides,* designed to kill fungus; and *rodenticides,* which kill rodents like mice and rats. It is a long-standing belief that pesticides are necessary for society to produce high crop yields, increase the longevity of wood structures, and reduce mowing and clearing costs on utility rights-of-way. These benefits of pesticide use, while giving the impression of control and security, overshadow many adverse long-term effects.

During the Second World War, DDT, the grandfather of pesticides, was used against malaria, typhus, cholera, and encephalitis by the U.S. Army.[1] In the post-war years of economic growth and development, synthetic organic chemical (SOC) pesticides, derived from inexpensive and abundant petroleum, became the crutch of world agriculture. The pesticide industry became a building block of the economy. Today, pesticide production is a multi-billion-dollar-a-year industry with an almost sacred status.

There are roughly 40,000 commercial products containing about 1,400 registered active ingredients. Of the approximately 600 active ingredients currently in production, 200 are considered major in terms of usage.[2] Pesticide consumption in the U.S. was an estimated 1.1 billion pounds in 1985,[3] up from a mere 464,000 pounds in 1951.[4] In 1987, about 429 million pounds of active ingredients were used on the major field crops.[5]

Considering the recent and ever increasing dependence on synthetic pesticides for a wide variety of uses, it is hard to convince some that man did manage to cultivate the land and sustain himself for thousands of years without them. Hundreds of years before Christ, the Chinese proved a mastery of pest control techniques by utilizing their natural enemies and optimum crop-planting times.[6] They utilized the hierarchy of life that exists in the natural world. In the last century, leading U.S. scientists supported management of habitat to take advantage of natural biological and environmental controls to deal with pests. Stephen A. Forbes, an entomologist at the University of Illinois, advocated the use of ecological methods in controlling crop insects.[7] And earlier this century, before the discovery of DDT, farmers utilized mechanical methods, biological controls, and less damaging inorganic pesticides to keep crop pests in check.

These natural and basic ideals and practices that assure sustainable and optimal crop production levels were forgotten with the advent of SOC pesticides and the quest for maximum yield. Pesticides helped farmers overcome the short-term cash flow dilema, but the long-term has proven to magnify these financial problems. Today, farmers are almost completely dependent on SOCs for controlling crop pests. Recent generations of farmers in the U.S. depend so heavily on pesticides that only a very small number have the knowledge to farm without them. In a reversal of a well known adage, the invention of synthetic pesticides has proven to be the mother of their necessity.

Synthetic chemical pesticides were an inevitable byproduct of an oil-based society. Their development was a predictable extension of increasing dependence on chemicals. Their initial low cost and apparent success was heralded as a revelation and began the pesticide age that has snowballed into today's staggering usage. Chemical companies convinced farmers and society of the necessity of pesticides and nature's job of controlling pests was taken over by synthetic chemicals.

In 1962, Rachel Carson's book, **Silent Spring,** alerted the world to the adverse effects of pesticides: destruction of non-target wildlife and natural pest predators; widespread contamination of water supplies; and untold human health effects. But, Carson's warnings were not heeded. New pesticides were synthesized and pesticide sales increased.

With the 1970s came a movement away from chlorinated hydrocarbon pesticides, the first generation pesticides that followed DDT, as many were banned or severely restricted. The year 1972, when DDT was banned, marked the beginning of the end for chlorinated hydrocarbon pesticides. Today organophosphate and carbamate classes of pesticides are predominantly in use. Organophosphates (made from nerve gas) and carbamates are cholinesterase inhibitors. Cholinesterase is a nerve enzyme which regulates nerve impulses. These pesticides restrict or inhibit cholinesterase and hence the transmission of impulses from nerves to muscles.

It was not until 1972 that meaningful legislation on pesticide regulation was enacted. The EPA's task under the Federal Insecticide Fungicide and Rodenticide Act (FIFRA) is to regulate pesticides by balancing the benefits against the risks (as currently perceived) associated with their use. This is done on an individual pesticide basis. New pesticides registered since 1977 are required to be tested for toxicity and other detrimental effects before registration and use is permitted. However, most pesticides currently in use were registered before the current guidelines were enacted. Toxicity effects data are either incomplete, inaccurate, or nonexistent for most of these pesticides. The result is a gaping hole in information on which to base regulation. Like many industrial SOC's, pesticides are the subject of regulatory decisions often based on ignorance and educated guesswork.

Health Effects of Pesticides

Several detrimental health effects have been associated with pesticide exposure: cancer; birth defects; infertility; Parkinson's disease; and other genetic, reproductive, and central nervous system abnormalities.[8] Pesticides are unique SOCs in that they are specifically designed to kill a living system and are deliberately spread over the land. It is irrational and grossly ignorant to think, as some do, that a chemical can kill a specific living organism while having absolutely no ill effects whatsoever on other living systems like humans.

Concern also exists for the undiscovered dangers of pesticides. A 1983 staff report of the Subcommittee on Department Operations, Research, and Foreign Agriculture of the U.S. House of Representatives Committee on Agriculture estimated that of the federally registered pesticides, 48% do not have the data needed to assess their potential to cause tumor growth, 38% do not have birth defects data, 48% lack data on reproductive effects, and 90% do not have mutagenicity information.[9] In 1984, the National Research Council of the National Academy of Sciences (NAS) released a report finding that health effects data is complete for only 10 of more than 3,350 pesticide active and inert ingredients.[10] Also, according to the U.S. General Accounting Office (GAO), given contemporary resources and the EPA's task of evaluating some 600 pesticide active ingredients, the completion of reregistration of all current pesticide products, with respect to up-to-date health effects data, will be sometime in the 21st century.[11] The GAO has concluded that:

> "the American consumer may not be adequately protected from the potential hazards of pesticide use because of the unavailability of information on pesticides to which much of the population is exposed daily."[12]

Cost/benefit analysis is the basis for pesticide regulation. The benefit side is placed squarely on the consideration of whether society will suffer economically if the pesticide is not used. Cost is measured in terms of adverse health effects. It is assumed that there is some risk with pesticide usage. The extent of the risk is variable with each pesticide, and in many cases may not be fully known due to a lack of testing data. Pesticides must not cause an "unreasonable risk" while being economically and socially beneficial. Determining what is reasonable is clearly a debatable point. Balancing health effects and commercial interests in a world where agriculture and society itself has become addicted to the fast-remedy pesticide is proving a difficult task.

Many question the wisdom of allowing risk to humans and the environment when alternatives to heavy pesticide usage exist. One can also question whether the benefits are solely reaped by the pesticide manufacturer, while the cost in terms of health effects and suffering is shouldered by society. The EPA's Economic Analysis Branch states:

> "While pesticide producers, users, and consumers benefit from the use of pesticides, costs are distributed throughout the population in terms of acute and chronic toxic effects such as cancer."[13]

Farming and Pesticides

> *"And I brought you into a plentiful country, to eat the fruit thereof and the goodness thereof, but when ye entered, ye defiled my land, and made mine heritage an abomination."*
> —Jeremiah 2:7

Prior to World War II, farming methods were chiefly mechanical. The machine age rendered farmers capable of larger scale cultivation and greater crop production than was experienced in the last century. With the coming of SOC broad-spectrum pesticides, farming changed radically. Herbicides replaced mechanized or labor-intensive methods to control weeds, and now are applied to over 90% of row crop acreage in the U.S.[14]

All species have natural enemies that control their population. Weeds and insect pests are no exception. A wide variety of predator/prey relationships exist among insect and plant species. Undisturbed natural conditions do not allow one species to increase to overwhelming numbers. The problem is that chemical pesticides decimate beneficial insects as well as the target pests. So, the surviving pest population, without an adequate natural enemy population, skyrockets to overwhelming numbers. This is why there is a problem with crop damage from certain insects even though pesticides are heavily utilized. In addition, new pests that had not been a problem before pesticide usage, develop as a result of regional disruption of natural predator/prey relationships. Consequently, new pesticides must be developed to control them. Also, the resistance to pesticides developed by certain pests over time leads to the usage of more and more chemicals. This has led to what is now called the "pesticide treadmill."

Current federal programs which subsidize farmers by guaranteeing prices for some crops, coupled with incentives to leave parts of their land fallow in attempts to reduce surpluses, encourages heavy pesticide usage. A farmer can use heavy doses of pesticides and fertilizer to maximize yield on the acreage in use. In many cases, this results in increased pesticide runoff to waterways.

Another problem is the current food production system in the United States encourages farmers to plant large fields with only one crop. The resulting absence of a diverse variety of plant, insect, and animal populations in one large area only invites pests because there are no predator species to control them. So the farmer uses insecticides and herbicides to improve his crop yield. The idea of planting two or three different crops in the same field is totally foreign to most farmers, but it creates a more harmonious and natural balance within that ecosystem which would in turn reduce pest and weed damage. Reduced tillage methods used by farmers today also invite larger insect populations and plant disease. This makes it necessary for farmers to use 30% more pesticides than previous tillage methods.[15]

Crop losses in the last 30 years have stayed fairly constant despite a tenfold increase in pesticide usage. In California alone, pesticide usage has increased threefold in the last 20 years, with farmers paying $1 billion yearly, while pest damage is **up** 40%.[16] Adding further to the modern farmer's troubles is the resistance to pesticides many insects are exhibiting. Today, over 445 insects are known to be resistant.[17] Pesticide resistance increases overhead for farmers through crop loss and higher application costs. The tobacco budworm made growing cotton in Texas unprofitable.[18] As it happens, the tobacco budworm is an example of a **secondary pest.** It only became a problem when its population increased dramatically due to the elimination of its natural enemies by broad-spectrum pesticides.

The quest to maximize crop yield each year has been the catalyst for the chemical industry's sales pitch. They encourage farmers to use expensive pesticides as well as large amounts of fertilizer. However, this strategy has proved costly, financially as well as environmentally. Between 1972 and 1980 despite an interest by a few in non-chemical "organic" farming, pesticide use increased 75%, while the cost to American farmers went up 400%.[19] This in turn caused pesticides to be applied more heavily with a greater chance for contamination of water supplies.

Non-Agricultural Pesticides

Surprisingly, about 50% of pesticide usage is in non-agricultural capacities such as building materials (wood preservatives), food containers, school yards, road sides, golf courses, parks, schools and public buildings (fumigants), railroad beds, and utility rights-of-way.[20] Over 60-million acres of rights-of-way, roadsides, electric, telephone, railway lines, and firebreaks, are routinely treated with herbicides like 2,4-D and 2,4,5-T to eliminate vegetation.[21] Some other non-agricultural uses of pesticides are listed in Table 13-1.

Home and garden pesticides are another potential source of exposure. According to the National Academy of Science, there is more pesticide use in urban and suburban areas than on farms.[22] A survey of suburban Chicago by an environmental group found lawn treatment by pesticides averaged ten pounds per acre annually. Soybean farmers use about two pounds per acre.[23]

Elevated levels of certain home-use pesticides have been discovered in indoor air.[24] Some, applied outside the home, found their way into the home through routes like cracks in the foundation. A study of home pesticide use found that routine spraying increases the risk of leukemia in children living there. Regular use of garden sprays elevates the risk of leukemia in children in the household over six times what it would be if the sprays were not used.[25] Birth defects were implicated in a study of heavy pesticide use in Florida.[26]

Lawn care is a 1.5-billion dollar-a-year industry.[27] An estimated 700,000 Massachusetts homeowners spend over $120 million yearly for weedless lawns, according to the Boston Globe.[28] Literally millions of households nationally hire professional lawn care services every year. This is amazing considering the lawn care business did not even exist 20 years ago. Even more surprising is the fact that the lawn care industry is growing at a rate of 25% annually. Chemlawn, just one lawn care company, increased sales from $87- million in 1979 to $284-million in 1987.[29]

Currently, the EPA does not require chronic toxicity testing for many pesticides that are used exclusively for non-agricultural purposes. The GAO reviewed the EPA's data on chronic toxicity for 50 non-agricultural pesticides and found preliminary data for only 18. Of these, 17 lack the chronic toxicology data to complete the assessments.[30] Of the approximately 600 pesticide active ingredients, only 194 have *preliminary* assessments.[31] According to the GAO, the EPA considers exposure to be insignificant because of the generally low levels of chemicals in the non-agriculatural products. They will not require chronic toxicity testing unless information exists to indicate widespread human or environmental exposure to these chemicals.[32] At the same time, however, the GAO points out that the EPA does not have the exposure information it needs to accurately and reliably assess risks for non-agricultural use pesticides.[33]

Many people who purchase and apply pesticides in and around their homes, or hire professional applicators, have absolutely no concept of the risks and possible health effects involved. They simply do not equate "that spray stuff" with something that is hazardous. The average American consumer is ignorant of the dangers lurking on the shelves of stores selling home use pesticides. Many assume that some all-knowing power somewhere would not allow it to be sold or used if it were not safe. Labels on pesticide products sold over-the-counter do not mention that the contents have been linked to chronic health effects in humans or lack adequate toxicity testing. The consumer carries a false sense of security along with the pesticide when he walks out of the store.

Often people are not even aware that the product they are using is a pesticide. Advertising the product as a spray, or marketing the product under a cure-all or quick-fix brand name may result in the illusion of safety in the mind of the user. Many products routinely used around the home are actually pesticides. These include:

- mildew sprays;
- algae control in swimming pools;

Table 13-1
Some Places Where Non-Agricultural Pesticides are Used

Parks and recreation areas	Hospitals
Restaurants	Mass transit stations
Department stores	Wood processing facilities (saw
Hotels	mills, etc.)
Industrial work areas	Health facilities
Retail food stores	Swimming pools and hot tubs
Tennis courts	Highways
Office buildings	Utility rights-of-way
Airplanes	Railroad rights-of-way
Sports arenas	Paper and textile mills

Source: **EPA Journal,** May, 1987.

- flea powders and dips for pets;
- mothballs (naphathaline);
- insect repellants; and
- weed, garden, and other outdoor sprays.

Some pesticide manufacturers even make claims about the safety of their products that the EPA considers "false or misleading."[34] Many professional applicators use EPA registration as an implication of safety, and the literature supplied by some manufacturers stresses the low human toxicity and safety of their product.[35] Simply because a pesticide product is registered with the EPA does not mean that it is safe. However, the complete safety of registered pesticides is a myth believed by a large portion of the population.

Misuse of pesticides by homeowners is also a problem. Many people assume that when applying a pesticide, more is better. This can result in increased exposure levels. Since many pesticides used to control home pests are soil or foundation injected, application of large amounts have a strong potential for contamination of well water.

Alternatives to home and garden pesticide usage do exist. Balancing cancer and birth defects against the psychological euphoria of a dandelion-free lawn or weed-free garden seems at the very least pointless, and at the most, ridiculous. Lawn care fanatics can still have nice lawns by proper aeration and fertilization techniques. Organic gardening methods have proven as effective or more effective than current farming methods on any scale. Good housekeeping and non-pesticide methods will keep unwanted pests out of the home. Pesticides with non-agricultural uses are listed in Table 13-2.

Table 13-2
Pesticides with Non-Agricultural Uses
(listed in descending order of quantity used)

Chemical	Type of Pesticide	Chemical	Type of Pesticide
2,4-D*	herbicide	polybutene	rodenticide herbicide
chlordane*	insecticide	trichlorfon	insecticide
sulfuryl fluoride	insecticide	endothall, dipotassium salt of	herbicide
diazinon*	insecticide		
chlorpyrifos (dursban)	insecticide	aspon	insecticide
betasan	herbicide	diquat dibromide	herbicide
heptachlor	insecticide	benefin	herbicide
atrazine	herbicide	piperonyl butoxide	insecticide
dacthal (DCPA)	herbicide	glyphosate	herbicide
carbaryl*	insecticide	lindane	insecticide
methoxychlor*	insecticide	acephate	insecticide
aldrin	insecticide	pentachlorophenol	fungicide
malathion*	insecticide	copper sulfate pentahydrate	fungicide
diuron	herbicide		
bromacil (Hyvar X)	herbicide	phorate	insecticide
sodium metaborate	herbicide	boric acid	insecticide
sodium chlorate	herbicide	picloram	herbicide
dichlorvos (DDVP)	insecticide	safrotin	insecticide
simazine	herbicide	ferric sulfate	herbicide
bendiocarb	insecticide	baygon (propoxur)	insecticide
parathion	insecticide	tebuthiuron	herbicide
dimethylamine dicamba	herbicide	chlorothalonil	fungicide
metolachlor	herbicide	maneb*	fungicide
dicofol	insecticide	captan*	fungicide
prometon	herbicide		
alachlor	herbicide		

*Ten most common pesticides used by homeowners.

Source: U.S. GAO and U.S. EPA.

Inert Ingredients

Only within the past few years has the EPA begun to address the problem of *inert ingredients* in pesticide formulations. An *inert ingredient* is the "carrier" of the pesticide *active ingredient,* and is not a factor in the efficiency of the pesticide. It can however, be a potential contaminant. The EPA lists 1,000–1,200 inerts used in approximately 49,000 pesticides. Between 800 and 900 of these have insufficient health and environmental data to determine their toxicity. Fifty-five are of immediate concern based on their toxicity.[36] These include: benzene, carbon tetrachloride, chlorobenzene, chloroform, dioxane (not dioxin), ethylene dichloride, methylene chloride, 1,1,1,-Trichloroethane, perchloroethylene, formaldehyde, isophorone, xylene, glycol ethers, epichlorihydrin, and vinyl chloride.[37] Even known active ingredients like pentachlorophenol have been used as inerts. Dicofal, a home use pesticide was found to contain about 10% DDT as an "inert" component.[38] Currently, it is not required that inert ingredients be listed on pesticide product labels, even though they comprise a large percentage of the product. Their environmental fate has not been adequately explored.

PESTICIDES IN GROUNDWATER

A number of changes since the mid 1940s have lead to greater potential for pesticide contamination of groundwater. Most important is the dramatic increase in pesticide usage. Other factors include alternative cultivation methods and increased irrigation, allowing less runoff to surface water but more water filtration through the soil.

It was not until the 1980s that it was discovered that pesticides, as well as other contaminants, deliberately spread on the land surface could contaminate groundwater. It was thought that soil layers would filter out pesticides and prevent pollution of subsurface waterways. However, numerous reports of contamination resulting from the widespread use of pesticides have brought about the realization that pesticide use can be a major contributor to groundwater pollution. The effects of agricultural chemicals on groundwater quality is rapidly emerging as a significant environmental issue.

As with groundwater pollution in general, the extent of pesticide pollution of underground waterways is unknown. The testing that has been done was confined to specific areas with only a limited number of pesticides monitored in each survey. At last count, 74 different pesticides have been found in groundwater wells in 38 states due to agricultural application (Figure 13-1 and Table 13-3).[39] On the order of 50 pesticides

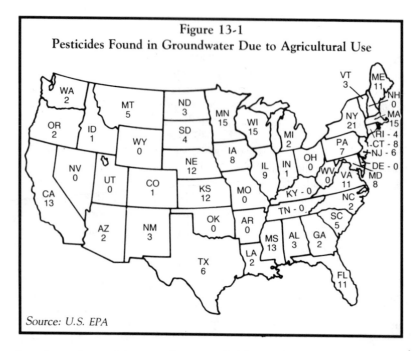

Figure 13-1
Pesticides Found in Groundwater Due to Agricultural Use

Source: U.S. EPA

have been detected in groundwater in California due to accidental spills, leaks, and direct land application.[40] Due to the lack of comprehensive monitoring for specific pesticides, little data exists on where contamination has or could happen. Consequently, predicting the level of human exposure to pesticides in drinking water, and the potential health effects, is very difficult.

The potential for groundwater contamination due to agricultural chemicals may be region-specific and not of national proportions. According to researchers at the Economic Research Service of the U.S. Department of Agriculture, the regions believed to have the greatest potential for pesticide contamination are the southern Coastal Plain, the Central Atlantic region, the Mississippi delta, the Midwest corn belt, Western Kentucky, and California's central valley.[41] Crops specific to these regions are pesticide-intensive crops. They include: corn and soybeans in the midwest; tobacco, cotton, rice, and peanuts in the southeast; and fruits and vegetables in California, parts of the northeast, and the Great Lake states. These crops generally have pesticides heavily applied to them.

According to several studies, the farming areas of the midwest have the highest rates of contamination. In Minnesota, about 40% of the wells

Table 13-3
Confirmed Pesticides Detected in Groundwater
Due to Agricultural & Point-Source Contamination

Arizona — DBCP

Arkansas — Alachlor, Atrazine, Metolachlor

California — 1,2-D, 1,3-D, Alachlor, Aldicarb, Atrazine, Bromacil, Chlorothalonil, DBCP, Diazinon, Diuron, EDB, Endrin, Simazine

Colorado — Atrazine

Connecticut — 1,2-D, 2,4-D, Alachlor, Atrazine, Dicamba, EDB, Metolachlor, Simazine

Florida — Alachlor, Aldicarb, Bromacil

Georgia — EDB

Hawaii — 1,2-D

Idaho — 2,4-D, BHC, Chlordane, DDT, Dicamba, Dieldrin, Heptachlor, Lindane, Malathion, Methyl, PCNB, PCP, Parathion, Silvex, Triallate

Illinois — Alachlor, Atrazine, Metolachlor, Metribuzin

Iowa — Alachlor, Atrazine, Cyanazine, Fonofos, Metolachlor, Metribuzin, Sulprofos

Kansas — 2,4,5-T, 2,4-D, Alachlor, Atrazine, Bromocil, Chlordane, Dieldrin, Endosulfan, Heptachlor, Metolachlor, Metribuzin, Picloram, Propazine, Trifluralin

Louisianna — Alachlor, Cyanazine

Maine — Alachlor, Atrazine Chlorothalonil, Dicamba, Dinoseb, Endosulfan, Hexazinone, Methamidophos, Peciloram

Maryland — Alachlor, Atraton, Atrazine, Cyanazine, Simazine, Trifluralin

Massachusetts — 1,2-D, Alachlor, Aldicarb, Carbofuran, Dinoseb, EDB, Oxamyl

Minnesota — 2,4,5-T, 2,4-D, Alachlor, Aldicarb, Cyanazine, Dicamba, MCPA, Metolachlor, Metribuzine, PCP, Picloram, Propachlor, Simazine

Mississippi — 2,4-D, Aldrin, BHC, Chlordane, DDT, Diazinon, Lindane, Malathion, Methyl parathion, PCP, TDE, toxaphene, trifluralin

Nebraska — Alachlor, Atrazine, Cyanazine, Dieldrin, Fonofos, Propazine, Simazine, Trifluralin

New Jersey — Atrazine, Chlorothalonil, DDT, Dieldri, Lindane, Simazine

New York — 1,2-D, 1,3-D, Alachlor, Aldicarb, Atrazine, Carbofuran, Cyanazine, Dacthal, Dinoseb, EDB, Ethoprop, Methomyl, Metolachlor, Oxamyl

North Carolina — Aldicarb

North Dakota — Parathion, Picloram, Trifluralin

continued

Table 13-3 (cont'd)

Pennsylvania — Alachlor, Atrazine, Cyanazine, Metolachlor, Propazine, Simazine

Rhode Island — Aldicarb, Carbofuran, Oxamyl

South Carolina — Aldrin, DBCP, DDT, Lindane

Texas — 2,4,5-T, 2,4-DB, Arsenic, Atrazine, Dicamba, Metolachlor, Prometan

Vermont — Atrazine, Cyanazine, Metolachlor, Simazine

Washington — EDB

Wisconsin — Alachlor, Aldicarb, Atrazine, Linuron, Metolachlor, Metribuzan, Picloram

Source: U.S. EPA

tested contained at least one pesticide.[42] Twenty-eight out of 70 public drinking wells tested in Iowa were found to contain one or more pesticides in measurable levels. Between 70% and 80% of all wells and springs tested in northeast Iowa by the Iowa Geological Survey had pesticides in them.[43] The researchers estimate that as many as half of Iowa's shallow wells may have low concentrations of pesticides.

Groundwater contamination by pesticides and nitrates may be a concern to 53.8 million people.[44] Dense populations and high dependence on private wells for drinking water in heavy pesticide usage areas are the reasons for this high number. Those on private wells are believed to be at a greater risk than people on public groundwater supplies because private wells are often shallower and therefore more vulnerable to contamination. They are also much less likely to be tested for pesticides, if they are tested for any contaminants at all.

To gain a better understanding of the extent of pesticide contamination of groundwater, the EPA is undertaking a nationwide survey. It is scheduled for completion in 1990. The EPA is testing 1,350 community and domestic wells for 60 pesticides that the EPA feels have the most potential to leach through the soil and contamintate groundwater (Table 13-4).

How Pesticides Contaminate Groundwater

In addition to leaching through the soil during routine agricultural application, contamination of groundwater by pesticides occurs through spills or leaks at manufacturing and storage facilities. This type of contami-

Table 13-4
Pesticides Considered to be Potential Leachers by the EPA

acifluoren	DCPA/dacthal	metolachlor
alachlor	diazinon	metribuzin
aldicarb	dicamba	nabam
ametryn	2,4-D	nitrates
ammonium sulfamate	1,2-Dichloropropane	oxamyl
atrazine	dieldrin	paraquat
baygon	dimethipin (b)	PCNB
bentazon	dinoseb	PCP
bromacil	diphenamid	picloram
butylate	disulfoton	prometone
carbaryl	diuron	pronamide
carbofuran	EDB	propazine
carboxin	fenamiphos	prophan
chloramben	fluometuron	sinazine
chlordane	fonofos	treflan
chlorothanlonil	hexazinone	triallate
cyanazine	maleic hydrazide	2,4,5-T
cycloate	MCPA	2,4,5-TP
dalapon	methomyl	tebuthiuron
DBCP	methyl parathion	terbacil

Source: U.S. EPA

nation is localized and classified as point-source (eminating from one point). Spills occurring during handling, or incorrect disposal of remaining pesticides after application, can result in **plugs.** Plugs are large amounts of heavily concentrated pesticide as opposed to lower less concentrated amounts that would occur through routine agricultural application over a wide area. Plugs overload the breakdown capacity of the soil, increasing the chances of the pesticide reaching groundwater at a larger concentration.

More commonly, though, pesticide contamination of waterways occurs through widespread land application to control insects and weeds on farmland, highways, homes and gardens, and utility rights-of-way. Pesticide application over these large areas creates the opportunity for runoff to surface water during rainfall, or migration to groundwater sources.

It is very difficult to predict in all cases whether a specific pesticide will reach groundwater. Several variables contribute to a specific pesticide's potential to contaminate groundwater. They are:

1. **How well the pesticide dissolves in water.** Water solubility is measured in ppm, just as pollutant levels are. A pesticide with a water solubility of 30 ppm or more is classified by the EPA as a possible leacher. Aldicarb, for example, has a water solubility of at least 6,000 ppm, while DDT, like other chlorinated pesticides, is nearly insoluble in water at less than 1 ppm.

2. **How well the pesticide adsorbs to soil.** This is an assessment of the pesticide's ability to attach to soil particles. Those that do not favor clinging to soil particles will be candidates for groundwater contamination.

3. **How volatile a pesticide is.** Non-volatile pesticides tend to be more persistent because they are not carried away in the air. However, those that are volatile are often applied by injecting them into the soil and covering the soil with plastic to minimize loss to the air. This, of course, changes the game, increasing the potential for the pesticide to reach groundwater.

4. **The half-life of the pesticide in soil.** Microbiolgical organisms in soil will break down pesticides. The rate at which a pesticide will degrade in soil is determined by the specific pesticide's *half-life* — the amount of time it takes for one-half of the pesticide to degrade. For example, if a pesticide is measured at 100 ppm in the soil directly after application, and two weeks later the level is 50 ppm, then the half-life is two weeks. The shorter the half-life, the faster it degrades, and the less likely it is to contaminate groundwater. A pesticide will degrade much more rapidly in soil than in groundwater due to the high amount of microorganisms in soil relative to groundwater. Consequently, when a pesticide reaches groundwater it will be much more stable and persistent, and its half-life will be extended. Certain pesticides can also remain in the soil over the winter, and with the spring melt levels in groundwater can increase dramatically. Herbicide concentrations in Iowa groundwater have been seen as high as 200 ppb during spring runoff.[45] The EPA is most concerned about pesticides with half-lives greater than 2–3 weeks.

5. **Type of soil the pesticide travels through.** Leaching of pesticides is much easier and quicker through coarse sandy soil than in very fine, heavy clay. However, recent studies indicate that some organic chemicals can penetrate thick clays that repel water.[46]

6. **Distance to the groundwater.** The greater the distance to the water table, the less chance for contamination. If factors con-

tributing to easy breakdown and slow leaching are present and it is a long way down to the water table, chances are that the pesticide will degrade before it reaches the water.

7. **When and how the pesticide is applied.** The extent of leaching can be increased by local rainfall. Application around the time of heavy rains can increase the leachability of the pesticide. The large amount of water movement downward increases the pesticide's mobility and its chances of reaching groundwater before it can degrade. Injection of the pesticide into the soil is generally believed to have the greatest potential for contamination. Pesticides like Dibromochloropropane (DBCP), which are manufactured to eradicate pests in the soil, have a higher likelihood of contaminating groundwater.

Utilizing a well irrigation system to apply pesticides (called chemigation), if not properly monitored, can increase the chances of pollution. Improper system usage or equipment failure can result in backsiphon or flow of large amounts of the pesticide directly into the well. Equipment malfunction of this type can also lead to very high levels of the pesticide, undiluted by water, being spread on the crops. Also, overirrigation of a field can override soil breakdown capability and increase runoff water.

Health Effects of Pesticides in Groundwater

Concern about exposure to agricultural chemicals in drinking water is increasing in farming regions where there is heavy reliance on groundwater and contamination of groundwater sources is believed to be widespread. Rural areas are considered most susceptible to drinking water contamination from pesticides. The farm family depending on a well for its drinking water could be subjected to elevated pesticide levels and consequently threats to their health. Pesticides in groundwater due to agricultural usage are usually found at relatively low levels. The significance to health at these low levels is not known for certain, but state and federal officials generally agree that no level of pesticides in groundwater used for drinking is acceptable.

A number of studies have concluded that modern farming methods (i.e. pesticide usage) contribute to a greater risk of dying of certain types of cancer. Researchers have found an excess of non-Hodgkins lymphoma associated with the use of certain herbicides, specifically 2,4-D.[47] Farmers whose main farming years were after WWII — the time of heaviest pesticide use — are at the greatest risk.[48] Dr. Leon E. Burmeister, of the Department of Preventive Medicine and Environmental Health at the University of

Iowa, reviewed studies on the relationship between pesticide usage and cancer/leukemia incidence in midwest farmers. He found an association between certain rare malignancies and exposure to certain pesticides.[49] It is speculated that because lung cancer is low in farmers and the application season is short, that chronic exposure, possibly through drinking water, is the cause of health effects rather than the careless application of pesticides.[50] According to Burmeister, there is an association between the increased usage of chemicals in this century and the increase in certain cancers, and one avenue of exposure may be the runoff of pesticides into shallow farm wells.

Individual counties with high leukemia rates exist throughout the midwest, from North Dakota to Texas.[51] Nebraska and Iowa farmers have elevated leukemia mortality rates, possibly due to heavy pesticide usage associated with areas of high corn production.[52] Herbicides are used on 93% of corn crop acreage.[53]

TESTING DRINKING WATER FOR PESTICIDES

Regulation of pesticides in drinking water has been minimal. Under the Safe Drinking Water Act (SDWA) only six pesticides out of 600 active ingredients are regulated in drinking water systems: Lindane, methoxychlor, toxaphene, endrin, 2,4-D, and 2,4,5-T. Lindane, methoxychlor and 2,4-D are the only ones with any degree of usage today due to restrictions and cancellations.[54] Even these three have some restrictions. The EPA is considering regulating additional pesticides in drinking water (Table 13-5).

Regulations do little for those using private wells which are the most vulnerable to pesticides and other farm-related contamination. However, these new regulations are bringing more comprehensive pesticide methods on-line at laboratories. Those with private wells can take advantage of their availability.

Considering all the pesticides in use it is impossible to look for them all. The first step is to determine what pesticides are used near the water source. The geographic pattern of pesticide applications as well as the crop uses of specific pesticides will determine which might be utilized in the area.

Certain pesticides are used on specific target crops. Eighty-five percent of all herbicides and 70% of all insecticides are used on corn, cotton, soybeans, and wheat crops.[55] Forty-seven percent of all insecticides are applied to cotton.[56] Between 85% and 90% of soybean, cotton, rice, and corn crops have herbicides applied to them. Alachlor, atrazine, butylate,

Table 13-5
Additional Pesticides Under Consideration for
Regulation in Drinking Water

Pesticide	Pesticide Type	Primary Crop or Other Uses	500 Series Methods
alachlor	herbicide	corn & soybeans	505,507
aldicarb aldicarb sulfoxide aldicarb sulfane	insecticide	variety of crops to control henatodes, mites, & insects	531.1
atrazine	herbicide & plant growth regulator	corn & soybeans	505,507
carbofuran	insecticide	corn	531.1
chlordane*	insecticide	termites in homes	505,508
dibromochloro-propane*	insecticide	wide variety of crop uses incl. vegetables & commercial turf	504
1,2-Dichloro-propane	insecticide	wide variety of crop uses & also used widely as industrial solvent	502.1
EDB†	insecticide	soil fumigant for soybeans, cotton, peanuts, pineapples, & other fruits & vegetables	504
heptachlor* heptachlor epoxide	insecticide	termites & ants in homes; corn, garden, lawn & turf insects	505,508
pentachlorophenol	wood preservative	wood preservation, seed treatment	515

*banned   †cancelled for most uses

Sources: U.S. EPA, **50 Federal Register,** No. 219, 13 November 1985; U.S. EPA, **54 Federal Register,** No. 97, 22 May 1989.

and metolachlor comprised more than half of all herbicides used on major crops in 1982.[57] Table 13-6 lists the most common pesticides used on major U.S. crops.

Some states place further restrictions on pesticide usage for specific crops beyond federal regulations. A call or letter to the state Agricultural Extension Service or state government pesticide branch could also help narrow the list of possible pesticides used. Also, usage of a particular pesticide may be geographically restricted by the marketing range of the company. One can even ask local farmers or applicators what they use, however, they may be reluctant to discuss their pesticide usage practices.

Table 13-6
Common Pesticides Used on Major U.S. Crops

Crop	Pesticides
Alfalfa	simazine
Barley & Oats	2,4-D, MCPA, triallate
Corn	carbofuran, chlorpyrifos alachlor, atrazine, cyanazine, dicamba, metolachlor, 2,4-D, butylate fonofus, simazine, terbufos, propachlor
Cotton	cyanazine, fluazifop-butyl, fluometuron, MSMA, norflurazon pendimethalin, prometryn, trifluralin
Flax	propachlor
Peanuts	alachlor, benefin, 2,4-D
Potatoes	alachlor, aldicarb, cyanazine, carbofuran
Rice	propanil, 2,4-D, molinte, 2,4,5-T, thiobencarb
Sorghum	alachlor, atrazine, propazine, 2,4-D, alachlor, metolachlor, propachlor
Soybeans	alachlor, aldicarb, bentazon, carbofuran, chloramben, chlorimuron-ethyl, ethalfluralin, fluazifop-butyl, metalochlor, sethoxydim, trifluralin, acifluorfen, lactofen, propachlor, metribuzin, imazaquin
Sugar Beets	aldicarb
Sunflowers	alachlor
Tobacco	diphenamid, isopropalin
Wheat	2,4-D, MCPA, chlorsulfuron, trifluralin, dicamba, triallate, bromoxymil

Source: Economic Research Service, USDA, **Agricultural Resources Outlook and Situation Reports 1988, 1987 and 1984; Agricultural Week,** Monday, October 27, 1987.

First Generation Pesticides

EPA 608, Organochlorine Pesticides (Table 13-7) is probably still the most common method for testing pesticides in water. A lab of any size will have this method on-line. Many of these pesticides can also be tested in 500 series methods (Table 13-7), however, they may not be widely available in many labs for a few years. How much EPA 608 will help is debatable. It is a test method that covers many of the first generation pesticides which were either banned or severely restricted in the 1970s. Many believe that the residues from the use of these pesticides are all but gone due to cancellations or severe restrictions placed on the majority of them. However, while they are not very water soluble, their use was widespread and they are very resistant to degradation. Chlordane and heptachlor were banned only recently for home use. Consequently, residues from these may persist for some time. The purchase of a home in farm country, or a home with a history of chlordane/heptachlor application and a well lying close to the house, may merit consideration of the test. One can only speculate on the past usage, spillage, or disposal of organochlorine pesticides in the area. Dumping or spillage of a large amount in a barn or storage area would cause an already persistent pesticide to persist even longer.

EPA 608 also covers the infamous PCBs. PCBs are not pesticides, but are similar chemically, thus they are easily analyzed by the same method. Their use, however, was completely different. The main use for PCBs was as coolants and insulation fluids in transformers and capacitors in the electrical industry.[58] Their resistance to chemical breakdown and stability at high temperatures made them well suited for industrial uses. However, these characteristics also make them persistent environmental contaminants.

PCBs were produced from 1929 until 1977 when Monsanto, the sole manufacturer, halted production. They do not degrade by natural processes, and will only break down at very high temperatures, which is why incineration is the preferred disposal method. PCBs do not dissolve in water, but are very soluble in the oil used in electrical components. The insolubility of PCBs makes them uncommon in high concentrations as water contaminants. However, those living near working or abandoned transformers, known areas of oil spills, or discarded electrical equipment might consider PCBs as potential contaminants.

Be aware that PCBs are grouped together commercially in mixtures called arochlors. Arochlors are numbered based on the molecular weight of the individual PCB components in the mixture. The significant majority of these arochlors found in water have been the 1242, 1254, and 1260 arochlors. Therefore, laboratory results are often expressed in terms of one or more of these three.

A method for testing the widely used herbicides, 2,4-D and 2,4,5-T has been available at many labs for some time. Although 2,4,5-T use is all but extinguished, 2,4-D remains one of the most extensively used pesticides. Both can be tested in EPA Method 515.1 (see Appendix IV).

New-Age Pesticides

The new-age pesticides, those in use today, are by comparison gener-

Table 13-7
EPA Method 608 for Pesticides and PCBs

Pesticide	Major Crops/Uses	500 Series Methods
aldrin	corn, structure protection, vegetable & fruit crops	505, 508
α BHC (Lindane) β BHC		505, 508
chlordane (mixture)	corn, potatoes, structures, lawn & turf, vegetables & fruits, ornamentals, shade trees, & livestock, cockroach & ant control in homes, wood infesting pests (banned in 1987)	505, 508
DDT DDE DDD	banned in 1972; widespread use in 1961, DDT was active ingredient in over 1200 formulations registered for use against 240 species of agricultural pests on 334 crops	508
Dieldrin	breakdown product of Termik, ants	505, 508
Endosulfan I Endosulfan II Endosulfan sulfate		508
Endrin	high mamalian toxicity, used for mice control in orchards, cotton, tobacco, & potatoes	508
Endrin Aldehyde		508
Heptachlor	fire ant control, closely related chemically to chlordane banned in 1969 except for home use completely banned in 1987	505, 508
Heptachlor Epoxide		505, 508
Toxaphene (mixture)	cotton, small grains, soybeans, corn, vegetables, fruit & nut crops, used to control grasshoppers & crickets in field crops	505
PCBs -1016 1221, 1232, 1242, 1248, 1254, 1260	PCBs are found in commercial mixtures called alochlors. (ex. alochlor 1248). The alochlors are listed at the left, the higher the number, the heavier the PCBs in that mixture. Alachlor 1260 is the heaviest.	505, 508

Source: U.S. EPA

Table 13-8
Common Pesticides and Testing Method Numbers

Pesticide	Method Number	Pesticide	Method Number
hexachlorobenzene	505,508	dicamba	515.1
dalapon	515.1	2,4,5-T	515.1
dinoseb	515.1	carbaryl	531.1
picloram	515.1	3-hydroxycarbofuran	531.1
oxamyl (vydate)	531.1	methomyl	531.1
simazine	505,507	butachlor	505,507
aldrin	505,508	metolachlor	505,507
dieldrin	505,508	propachlor	505,507
2,4-DB	515.1	metribuzin	507

Source: U.S. EPA, *54 Federal Register*, No. 97, 22 May 1989.

ally less persistent and degrade more rapidly than organochlorine pesticides. However, they remain toxic and are more soluble in water than their predecessors. This is part of the reason for concern about their existence in groundwater. In surface water, these pesticides are subject to attack by microorganisms and will degrade more easily. If they reach groundwater, where there is an absence of these organisms, they will degrade much more slowly. If it is a perpetual problem, such as routine application or undetected leakage, the contamination will be constant.

Many labs do not routinely analyze for many of the newer pesticides. However, approved EPA methods do exist for many common pesticides, and as new drinking water regulations are mandated for certain pesticides, more labs will have these methods on-line. Table 13-5 lists pesticides that will be regulated in drinking water along with corresponding 500 series methods. Table 13-8 lists additional common pesticides and method numbers. See appendix IV for complete 500 series pesticide methods. Some labs may have methods on-line to test for the pesticides commonly used in the state or area due to increasing consumer demand or state regulations. In some states, government environmental laboratories, or state health department labs may perform testing of groundwater for pesticides as part of surveys. They may perform analysis on samples from citizens for a fee, or they may be able to direct individuals to labs that can perform the service.

Table 13-9
Pesticides Reviewed or Under Special Review by EPA

Pesticide	Chronic Health Concerns*	EPA Regulatory Action
alachlor	oncogenicity	restricted to use by licensed applicators
aldicarb	acute toxicity	
aldrin	carcinogenicity	most uses cancelled
amitrole	carcinogen	
benomyl	reproductive effects	protective clothing required for applicators
captan	oncogenicity other chronic effects, mutagen-icity, tumors, birth defects	pending
carbaryl	oncogenicity teratogenicity mutagenicity	deferred (to be addressed in the reregistration process)
carbofuran	effects wildlife, bald eagles	
chlordane	oncogenicity	removed from market by manufacturers pending proof of no health risk if properly applied
chlorothalonil	tumors	
dichlorvos (DDVP)	oncogenicity reproductive effects mutagenicity	deferred (to be addressed in the reregistrtion process)
dicofol	none†	pending
dinocap	birth defects	
EBDCs	carcinogen, birth defects	
heptachlor	oncogenicity	removed from market by manufacturers pending proof of no health risk if properly applied
lindane	oncogenicity reproductive effects teratogenicity other chronic effects	some limited uses cancelled
linuron	carcinogen	
parathion	acute human toxicity	*continued*

Table 13-9 (cont'd)

Pesticide	Chronic Health Concerns*	EPA Regulatory Action
pentachloro-phenol	oncogenicity teratogenicity	non-wood uses: pending; wood uses: safeguards required for applicators
phosdrin	acute human toxicity	
piperonyl butoxide	oncogenicity	deferred (to be addressed in the reregistrtion process)
toxaphene	oncogenicity	many uses cancelled
trichlorfon	oncogenicity reproductive effects teratogenicity mutagenicity	deferred (to be addressed in the reregistrtion process)

*In addition to chronic health concerns, several chemicals presented environmental concerns.
†Stated concerns were for environmental effects.

Source: U.S. General Accounting Office, **Non-Agricultural Pesiticides: Risks and Regulation,** April 1986; U.S. Dept. of Agriculture, **Agricultural Resources,** January 1988.

As with other classes of chemicals, certain pesticides are related chemically and can be analyzed by the same method. So, when testing for pesticides, try to do so by a complete method. This covers more pesticides for the same amount of money. The number of tests run will determine how much testing will cost, not the number of pesticides tested.

Changing Attitudes Toward Pesticides

More and more people are waking up to the simple fact that pesticides are bad news. The health effects of an increasing number of them are being questioned and the uses of many are being restricted or cancelled (Table 13-9). Concern about the health effects of pesticides in food and drinking water are certainly changing people's attitudes toward pesticide usage practices. Some municipalities are restricitng lawn chemical usage, and organically grown fruits and vegetables are gaining popularity. The Davey Co., a lawn care company, has voluntarily cut back its use of pesticides by at least 75% as part of its Plant Health Care concept.[59]

Alternatives to the ecologically-damaging, drinking-water- contaminating, and financially-ruinous current farming methods do exist, although they are not widely accepted. Sustainable agriculture's goal is to restore and maintain nature's ecological balance. These methods minimize soil errosion which has ruined countless acres of farmland, reduce overuse of precious groundwater supplies which are becoming depeleted in the west, preclude the continuation of pest damage, and eliminate health effects from contaminated food and water. All these problems were created by the heavy dependence on chemicals. The idea behind sustainable agriculture is control of pests within the balance of the ecosystem rather than the irradication of both beneficial and pest organisms as is done by today's broad-spectrum pesticides. Sustainable farms revitalize and continually maintain the productivity of the soil while minimizing its errosion. Use of alternate farming methods like these on cotton resulted in corn replacing cotton as the crop with the largest amount of insecticide usage.[60]

The lower input costs of sustainable agriculture systems also favor the small or family farm. It reduces monopolization, which has the added advantage of diversifying rural economics, thereby cushioning them against recession. Large-scale chemical-intensive farming does not leave as much money in the rural farming community. Expenses paid for chemicals go to urban-based production centers.[61] In addition, organic farms reportedly use 50% to 63% less energy than conventional farms.[62]

In spite of the increased concern about the health and environmental effects of pesticides, proven ecologically sound alternatives are not being adequately explored by government, industry, and agriculture. It is unlikely that large chemical companies with a hearty investment in their SOC pesticide products will endorse natural biological control of pests. Once a viable farming ecosystem is in place, it is self-sustaining and no chemical products are needed. Sustainable agricultural methods have proven to promote a productive balance of life since man began to till the soil and can continue to do so without pesticides.

14. SOURCES OF CHEMICAL CONTAMINANTS

"Without a much more successful way of handling the risks associated with the creations of science, I fear we will have set up for ourselves a grim and unnecessary choice between the fruits of advanced technology and the blessings of democracy."

— William D. Ruckelshaus, U.S. EPA Administrator
in a speech before the National Academy
of Science Washington, DC, June 22, 1983

It is helpful to be aware of the potential sources of chemical pollutants when assessing water quality and deciding on which chemical tests to choose. This chapter discusses some of the most common pollutant sources in terms of identifiable disposal sites. The main focus will be on groundwater although surface water can be subjected to some of the same sources.

There are over 292 million tons of hazardous waste generated in the U.S. each year, over a ton for each person in the country. The sources of chemicals that have the potential to pollute groundwater, and consequently well water, are as varied and infinite as the ways of generating, storing, transporting, and disposing of chemicals. However, some sources are more common than others. As one might expect, the types of contamination in a particular region often reflect the human activities common to that area, such as mining, farming, or specific chemical industries.

Injection Wells

Currently, underground injection is the principle disposal method for hazardous waste.[1] Wells are used to pump liquid waste into underground deposits below usable water sources where it will ideally remain forever (Figure 14-1). About 10 billion gallons are disposed of yearly, accounting for 58% of all waste disposed.[2] This is ten times the amount put into landfills and twice the amount put in surface impoundments.

There are about 252 wells in the U.S. injecting what is classified as hazardous wastes.[3] This includes organic chemicals, metals, and corrosive wastes. The majority of these wells are concentrated in certain areas of the country with Texas and Louisianna accounting for 66%, and Ohio, Michigan, Indiana, Illinois, and Oklahoma with most of the remainder.[4]

There are also an estimated 140,000 brine-injection and oil-recovery wells associated with oil and gas drilling.[5] Brine is classified as saline or salty water high in sodium, chloride, other inorganics, metals, and hydrocarbons characteristic of deep aquifers that are in or near oil and gas formations. While the amount of brine waste injected will depend on the extent of oil production, the Office of Technology Assessment (OTA) estimates the amount of brine produced annually to be about 525 billion gallons. The majority of this is injected underground.[6] Estimates of the number of barrels of brine produced per barrel of oil range from four to 100.[7] Sewage disposal and agricultural and urban runoff water are handled by an estimated 40,000 additional disposal wells.[8]

The contamination of groundwater from underground injection of wastes can occur through the improper construction, operation, or closing of the well. In areas of oil and gas drilling, where most injection wells exist, there are many abandoned or poorly operated exploration and production wells. When brine or other waste is injected into a disposal well, it can move horizontally underground to these wells. If a production well was poorly constructed, has degraded over time, or was not adequately capped or plugged, the waste can travel vertically and leak into groundwater formations.[9] Contamination can also occur through injection directly into drinking water aquifers, by waste pushing saline water up into drinking water aquifers, or by the movement of waste into a drinkable water area of the same aquifer.[10]

Even though the practice of underground injection is widespread, the actual fate of injected hazardous waste is poorly understood. The risk of widespread groundwater contamination is clearly possible considering the huge volumes of waste that are injected into deep wells. Once in the ground, these wastes cannot be removed or controlled in any way. There

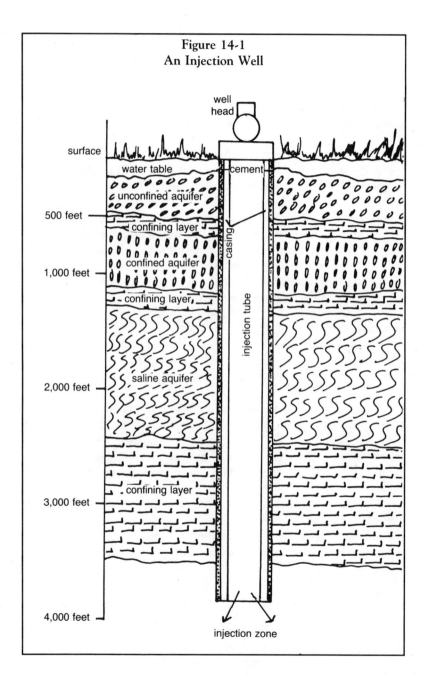

**Figure 14-1
An Injection Well**

is limited knowledge of the subsurface environment. Its inaccessibility makes it nearly impossible to accurately predict the eventual movement of the waste underground or to be certain the waste remains in the geologic formation into which it was injected.

The combining or mixing of reactive or incompatible wastes underground could produce gases or solids and cause pressure buildup. This could result in the fracturing or cracking of geologic formations, allowing the waste to migrate over great distances underground. Scientists believe waste injection in northeast Ohio and near Denver, Colorado resulted in an increase in earthquake activity these areas.[11] Deep well injection in Ontario caused a number of abandoned Michigan brine wells to begin flowing due to the significantly increased aquifer pressure.[12]

In spite of the uncertainty associated with underground injection, it is becoming more attractive as incentives for surface disposal decrease. It is less expensive than some other methods like incineration, and underground injection control (UIC) regulations are not as strict as Resource Conservation and Recovery Act (RCRA) regulations for waste disposal.[13]

Landfills

Landfills often arouse concern due to the possibility of contaminating nearby wells. The concern is well founded. Landfills and surface disposal in general, have often resulted in contamination of the drinking water of nearby residents. Some landfills that existed in the days of the Roman Empire still leak contaminated runoff. The exact number of landfills is not known, but estimates are that 18,500 municipal,[14] and almost 76,000 industrial sites exist in the United States.[15] Seventy-five thousand landfills are classified as on-site (an industrial property) and little is known about them.[16] Fewer than ten states require regular groundwater testing at these facilities.

Groundwater contamination by landfill runoff occurs because landfills were inadequately sited and constructed with respect to the area's geology and proximity to groundwater. Landfills were often situated on low value or unsuitable land like marshes, abandoned sand and gravel pits, old strip mines, or limestone sink holes. All are prime areas for groundwater contamination. Futhermore, the locations of many older landfills are not known. Many landfills also lack effective monitoring to determine if they are leaking.

The great majority of the more than 93,000 landfills in United States are not classified as receiving hazardous waste. However, concern exists for all landfills because of widespread chemical usage and the tradition of poor or nonexistent record-keeping at landfills. Much of the industrial manufacturing and consumer wastes containing synthetic organic chemicals end up in landfills.

Surface Impoundments

Surface impoundments are holding ponds, pits, or lagoons that store liquid wastes. These wastes are eventually either injected underground, discharged into waterways, simply allowed to evaporate, or are left to leach into the ground. The EPA conservatively estimates that a total of 181,000 surface impoundments exist across the country. They hold industrial, municipal, mining, and oil and gas drilling wastes.[17] Ninety-three percent are considered lacking in maximum protection of groundwater because of the geology existing beneath them or due to their close proximity to groundwater. Ninety-eight percent are within a mile of current or potential drinking water sources, and about 70% of the industrial impoundments are unlined.[18]

The potential for groundwater contamination from impoundments or holding ponds is believed to be quite significant. One report found 23,000 cases of surface and groundwater contamination in Texas alone due to leaking or seeping brine disposal pits.[19] A wide variety of hazardous wastes including SOCs, heavy metals, and high levels of inorganics are likely to exist in industrial impoundments. Reactions between different compounds in the waste may create even more compounds.

As with other waste sites like landfills, foresight and knowledge of potential problems were lacking when areas for waste disposal were selected. Criteria for picking a site for industrial impoundments was based on convenience and low cost as opposed to potential for groundwater contamination.

Septic Tanks

Since septic tank systems were first introduced in the U.S. in 1884, they have become the most widely used on-site method of sewage treatment and disposal.[20] Over 70-million people in 17-million homes, 1/3 of all households, use septic tanks for disposal of home wastewater.[21] According to the EPA, 25% of new homes constructed in this country use septic tank systems.[22]

To the surprise of many, septic tanks are the most often reported source of contamination of drinking water wells.[23] Some homeowners find their own septic tank is the source of their well contamination. About 800-billion gallons of wastewater are discharged from septic tanks into

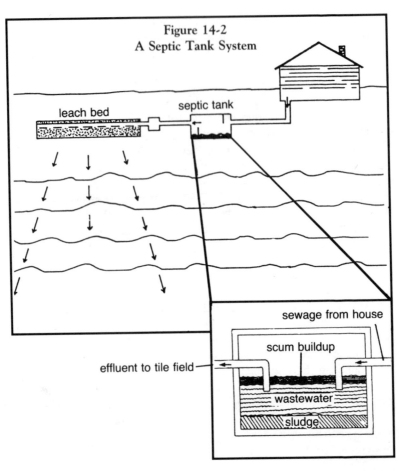

Figure 14-2
A Septic Tank System

leach bed

septic tank

sewage from house

scum buildup

effluent to tile field

wastewater

sludge

drain tile fields every year, delivering more wastewater straight into the soil than any other pollution source.[24] An estimated 40% of the septic tanks now in use do not function correctly.[25] Most were installed with new housing in the 1960s, and the life span of these systems, estimated at anywhere between 10 and 40 years,[26] is starting to run out.

A septic tank system consists of: 1) an underground tank which accepts wastewater directly from the house, where solids settle out and are broken down by bacteria; and 2) the drain field or leach bed, which distributes the liquid from the tank over an area of subsurface soil. Theoretically, the wastewater is either filtered by the soil or biodegraded by microorganisms before it reaches groundwater (Figure 14-2).

One of the major advantages of the septic tank system is its low maintenance. Pumping out sludge is necessary only about every 5 years.[27] Septic tanks are low cost, low technology, and require minimal energy compared to municipal treatment plants. In any case, the responsibility of assuring proper operation of the septic tank system and performing the required maintenance, however minimal, usually lies with the homeowner.

One of the chief problems associated with septic tanks in many areas is human population density. Too many septic tanks in a given area overload the natural purification ability of the soil and allow large volumes of wastewater to reach the watertable. The denser the population, the greater the likelihood of contamination. According to the EPA, a density of more than 40 domestic septic systems per square mile makes an area a prime target for subsurface contamination.[28] The more heavily populated areas of the east, and parts of California, fall into this category.

In addition to bacteria and viruses found in human waste, nitrates have also been contaminants of concern from septic tank drainage. High nitrate levels in groundwater on Long Island, New York, were found to originate from septic tanks and lawn fertilizers.[29] More recently, septic tanks have became recognized as a source of metals and synthetic organic chemicals (SOCs). Toxic SOCs are common in literally hundreds of products used in and around the home. In addition to contaminating the groundwater, some chemicals can corrode pipes thereby releasing even more heavy metals into the septic system. Many people, ignorant of the hazards, dump chemicals down the drain to their septic tank or simply dump them on the ground in their back yard. Household products that contain SOCs include:

- *pesticides* — garden sprays, household bug sprays;
- *paint and coating products* — oil based paints, wood preservatives and coatings, paint thinners, paint removers, varnishes, and waxes;
- *cleaners* — toilet cleaners, drain cleaners, disinfectants, degreasers, nail polish remover, stain removers, oven cleaners, dry cleaning fluids, air fresheners, shoe care products, rug cleaners, floor wax, metal polishes, and other laundry and cleaning solvents;
- *auto products* — antifreeze, engine degreasers, flushes for engines and radiators, transmission and brake fluids, used motor oil, gas, and batteries; and
- *other products* — kerosene, heating oil, rust proofers, refrigerants, tar and roofing products.

An EPA study of chemicals in septic tank effluent found toluene, methylene chloride, benzene, chloroform, and other volatile SOCs related to home chemical use. The study concluded that although volatiles removal is ordinarily a routine mechanism in septic tanks, little removal of the volatiles takes place in the tank if there is a thick scum layer on the surface of the tank's contents. They will then be discharged into the drain field where they could contaminate the groundwater.[30] Generally, however, the heavier the chemical, the less likely the chemical is effectively removed by the septic tank.

Some SOCs like trichloroethylene (TCE) and methylene chloride have even been used as septic tank and cesspool cleaners.[31] Cleaners dissolve the sludge and both the sludge-water and the cleaner are carried into the drain field. About 400,000 gallons of septic tank cleaner liquids that contained TCE, benzene, and methylene chloride were used by Long Island residents in one year![32]

Mine Tailings and Mine Waste Piles

Mining operations result in **waste piles** of soil or rock that are removed in the mining process, and **mine tailings,** which are the solid wastes that are left over after the ore is removed. These wastes are usually piled near the mining or smelting operations. Yearly, 2.3-billion tons of these waste tailings, including radioactive wastes, are generated during mining activities.[33] Rain or snow falling on tailings washes heavy metals and radioactive elements, such as uranium and radium, into surface water and groundwater.

Acid-mine drainage from mine wastes occurs when water and oxygen react with pyrite, a mineral found in mine wastes, to form a mixture of iron and sulfuric acid. This rust-colored water is found in coal mining regions of the eastern United States. In the west, the acid is neutralized by geologic formations, so acidic water is not as common. Metals like mercury, arsenic, zinc, and nickel are also associated with mine water.[34]

Underground Storage Tanks

According to the EPA there are between five and seven million storage tanks in the U.S.[35] This includes tanks of home heating oil, industrial petroleum products, solvents, and hazardous waste. An estimated 2.5 million underground storage tanks exist in the United States. The great majority of them are made of steel. Between 1.5 and 2 million are used by the petroleum industry, with 15% to 20% of these believed to be leaking. Service stations use about 1.2 million underground storage tanks. Farms, trucking companies, corporations, government agencies, and petrochemical companies also use underground storage tanks.[36]

Steel underground storage tanks are subject to rust and consequently leakage, a problem that was not considered when they were installed. A number of these tanks are quite old, installed in the 1950s during the expansion and development of the nation's interstate highway system. According to an EPA summary of leak reports, tanks begin to drain after an average of 17 years. Leaks of connective piping occur at an average of 11 years. The range, however, varies from under one to over 50 years.[37] In addition, an old abandoned underground storage tank with a small undetected leak can systematically seep into groundwater for years without being discovered.

No one knows how many abandoned underground storage tanks there are, or where they are located. Some have turned up in pretty surprising locations. A construction crew found a tank buried in the middle of Main Street in a Vermont town. In Lincoln, Nebraska, a leak was found during the installation of a sewer line, and was traced to 40 year old tanks in the basement of a bank. Apparently a Trix Oil gas station had existed previously on the bank site and the tanks were never removed.[38]

The EPA estimates that the number of tanks that may be leaking will likely increase to 75% of the total.[39] Several hundred thousand to millions of gallons of gasoline and chemicals leak from underground storage tanks yearly.[40] New York State believes 16,000 of its 83,000 actively used tanks leak. According to the Steel Tank Institute, an estimated 350,000 tanks containing gasoline will leak within five years.[41] This number does not include those containing other solvents or hazardous waste.

Unprotected steel underground storage tanks have been known to leak for a variety of reasons: improper tank and pipe installation; corrosion due to moisture in the soil and the tank; corrosion due to electrochemical reactions between the metal of the tank and the soil; and reactions between the two different metals of the tank and the pipe (Figure 14-3). The piping system from the tank may be even more susceptible to leaks than the tank itself, and protection from these leaks is just as important. The tank piping system may leak due to: corrosion; improper alignment; stress from temperature such as freezing or thawing; surface traffic load over the pipe; and settlement of the fill soil.

Since regulations were tightened on underground storage tanks, fiberglass tanks are generally installed today. Fiberglass, of course, is less susceptible to corrosion and leakage but can rupture if the water table rises and forces the tank upwards under pressure. However, fiberglass tanks are preferred for their durability.

With the change to unleaded gasoline, compounds like benzene,

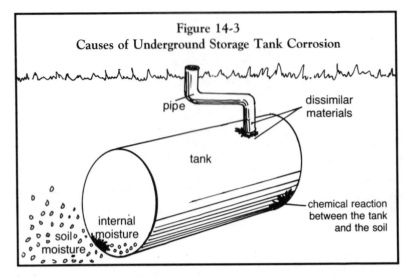

Figure 14-3
Causes of Underground Storage Tank Corrosion

pipe

dissimilar materials

tank

internal moisture

soil moisture

chemical reaction between the tank and the soil

toluene, xylenes and ethylbenzene have been added to gasoline to increase octane. These chemicals move more readily in groundwater than other gasoline hydrocarbons. Since they are volatile they will float on the water layer and mix only slightly with the groundwater. They can, however, be pulled into a well during pumping.

The costs of cleaning up of underground storage tank leakage, as one might expect, could be very expensive. Provincetown, Massachusetts spent over three million dollars in attempts to clean an aquifer contaminated by just one seeping tank.[42] The cost to clean all the leaking tanks could be astronomical.

Open Dumps

Indiscriminate dumping areas and open burning sites are classified as *open dumps*. The types of hazardous waste they contain can be as unique and varied as the imagination allows. People ignorant of the dangers of the chemicals they routinely use carelessly dispose of them in these locations. Precipitation washes chemical wastes from these dumps into surface water as well as groundwater (Figure 14-4). Open burning at these sites also leaves residues that can be washed into groundwater. Some commercial and industrial sites of open burning are now on the EPA National Priority List and groundwater has been contaminated at these sites.[43]

Highway Salting

From 1975 to November 1979, the town of Schodack, New York

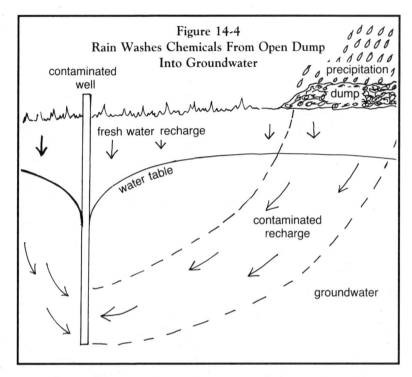

Figure 14-4
Rain Washes Chemicals From Open Dump Into Groundwater

contaminated well

precipitation

dump

fresh water recharge

water table

contaminated recharge

groundwater

stockpiled road-salt outside the town garage. In the fall of 1979 the road salt/sand mixture was dumped near the top of a 100 foot well. In November 1980, complications related to kidney disease in a local resident prompted an investigation that found high levels of chloride and bacteria in his well.[44] Salt and chloride are the chief contaminants associated with highway salting. Major contamination problems are usually associated with stockpiles at storage facilities where the concentration of contaminants is greatest. Runoff from a storage of salt resulted in the contamination of the water supply of three hundred Michigan families.[45] Chromate, added to reduce the corrosiveness of the salt, is believed to be the source of increased levels of hexavalent chromium in some waterways after the snowmelt.[46]

Animal Feedlots

Animal waste from feedlots in farming areas are a prime source of microbiological and nitrate contamination. The potential for contamination is highest where the water table is shallow and a large number of animals are confined in a small area.[47] Since animal feedlots are usually found near rural homes, drinking water wells can be at risk for contamination.

Atmospheric Deposition

Pollutants discharged into the air by industrial processes can deposit in surface water or on land where they may be washed into waterways. The most widely known example of this is acid rain, the deposition of acidic water caused by the mixture of sulfur and nitrogen oxides with rain. Many other chemicals discharged into the air can also contaminate surface and groundwaters. A factory in Michigan discharging dust containing chromium through roof ventilators was responsible for the accumulation of the dust just downwind from the plant. The compounds containing the chromium were very water soluble, and the groundwater supplying a municipality was contaminated.[48] In Ohio, the discharge of airborne fluoride at an aluminum processing plant resulted in fluoride levels in groundwater above 1,000 mg/L.[49]

Underground Pipelines

About 155,000 miles of pipeline carry hazardous waste in the United States.[50] Over 216,000 barrels of liquid waste were lost in reported accidents in 1988 and clean-up costs were in the tens-of-millions of dollars.[51] Line ruptures, corrosion, weld failure, operational mistakes, earthquakes, and root damage have been known to cause leaks.[52] Additionally, pipelines carry natural gas, sewage, and other products. Leaks in pipelines are often not detected until waterways or vegetation become noticeably affected. About 90,000 gallons of toluene from an underground pipeline spilled into the Sandusky River in western Ohio in February, 1988. Large fish kills were reported, thousands of people were evacuated, and companies were forced to shut down to conserve water.[53]

Transport of Hazardous Waste

According to the EPA, an estimated 90% of all hazardous waste is transported by truck. They also estimate that for every 11,000 gallons shipped, 38 are lost in the shipment or transfer of the waste. The OTA estimates total nationwide loss to be about 14 million tons annually.[54] The amount that affects groundwater is not known. This does not include contaminants classified as non-hazardous.

"Midnight Dumping"

Illegal dumping by waste generators to avoid disposal fees is a serious problem, and will continue to be as long as costs of required disposal are on the rise. According to a General Accounting Office (GAO) report, the actual extent of illegal waste disposal is unknown.[55] In an attempt to characterize the amount of illegal disposal, a survey by an EPA consultant estimated that one in seven hazardous waste generators had illegally disposed of wastes within a two year time period. [56]

Under the Resource Conservation and Recovery Act (RCRA) hazardous waste generators are held responsible for the proper disposal of their waste. The waste is followed from "cradle-to-grave" by a **manifest** which is passed on from the generator to the transporter and ultimately to the disposal facility (Figure 14-5). A signed copy of the manifest is then returned to the generator. An unscrupulous transporter, given the manifest and paid to transport and dispose of the hazardous waste by the generator, can forge the disposal facility's copy of the manifest and return it to the generator. The transporter can then illegally dump the waste and pocket the disposal costs paid to them by the generator.

Verification of the volumes and types of hazardous waste reported by generators requires considerable cost and effort by regulatory agencies.[57] Proper disposal of hazardous waste is generally on the honor system. In addition, the EPA cannot assure correct disposal by all operators because they are not aware of all generators and may not have complete data on the amounts and types of waste produced.[58] Those waste generators that the EPA is unaware of, those working outside the law, have free reign. They have not notified the EPA that they are involved in activities which result in hazardous waste generation and can either dispose of waste on-site without a permit, or dump it somewhere off-site without a manifest. Dumping the waste down the drain into sewage systems is not easily detected. Analysis of hazardous synthetic organic chemicals is not generally done at sewage treatment plants.

Off-site dumping, such as roadside dumping or disposal in secluded areas, is a common illegal method of hazardous waste disposal. Water in roadside ditches is infrequently, if ever, tested. Even if it is tested and contamination is found, it would be nearly impossible to find the party responsible. Unlike specific sites of hazardous waste disposal like landfills, illegal dumping areas are unknown. They can be anywhere. "Midnight dumping" can involve large amounts of waste dumped directly on the ground within a small area. This would likely result in high levels of contaminants in groundwater and drinking water wells.

Illegal dumpers have used some very innovative methods to avoid detection. One tanker truck driver was caught dumping brine-waste while

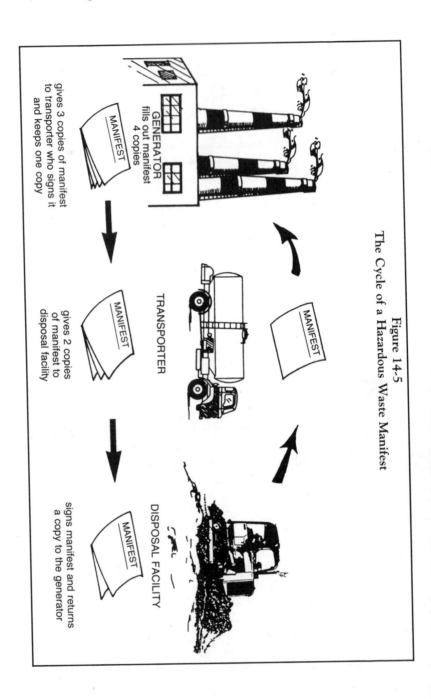

Figure 14-5
The Cycle of a Hazardous Waste Manifest

GENERATOR
fills out manifest
4 copies

gives 3 copies of manifest
to transporter who signs it
and keeps one copy

MANIFEST

TRANSPORTER

MANIFEST

gives 2 copies
of manifest to
disposal facility

MANIFEST

DISPOSAL FACILITY

signs manifest and returns
a copy to the generator

MANIFEST

driving down the highway in a rain storm with the valves open. Those involved in illegal disposal also make it difficult to be identified. A textbook example of illegal waste disposal is a tanker truck sitting by the roadside looking like it is supposed to be there, perhaps even with flashing amber lights. It has a hose running from the truck into a roadside ditch or waterway, but there will be no identifying marks on the truck; no company name, no registration number, not even a license plate.

You can actively participate in curbing this type of disposal. In fact, according to a GAO survey of enforcement actions, in 34 out of 36 cases, a tip by a citizen or employee resulted in the detection of the illegal disposal activity.[59] California and New Jersey even have monetary awards for people who provide information that leads to the arrest and conviction of illegal waste dumpers.

15.
INEXPENSIVE SCREENING TESTS
Getting the Most for Your Money

Many water quality tests can be expensive. In particular those for synthetic organic chemicals (SOCs) are quite costly and, depending on the method, may cover only a small number of chemicals. However, *screening* or *indicator* tests generally cost less because they do not require the highly specialized technology needed to identify specific pollutants. Although some of these methods do not measure the amount of a specific chemical, in a way they go beyond this. A screen or indicator test often encompasses a large group of chemicals or provides a great deal of information about a facet of water quality. Some cover more territory than specific testing methods and, therefore, might be considered more beneficial in assessing general water quality.

This chapter will cover screening methods used in analyzing for SOCs in drinking water. The two most common indicator tests for organics are Total Organic Halide (TOX) and Total Organic Carbon (TOC). Both give good overall information about water quality rapidly at a relatively low cost. TOX will determine the level of an entire class of chemicals in your water, while TOC will give an indication of the stress that biological or chemical pollutants are placing on the water system.

If the cost of comprehensive analysis for specific pollutants is out of reach financially, indicator methods are the best alternative. Large companies when monitoring their own effluents, and federal and state agencies when determining landfill or hazardous waste leakage, will use these indicator tests to narrow down the possible contaminants before continuing with more expensive, complicated, and time-consuming methods. TOC and TOX are used to test wells, landfills, and other waste sites to determine the spread of contamination in groundwater. The area of a plume or zones of groundwater contamination can be mapped inexpensively by using these indicator tests. The consumer, however, only wants to know whether his water is safe or not. Screening tests have the advantage of giving a "thumbs up" or "thumbs down" on water quality with respect to a large group of pollutants.

TOTAL ORGANIC HALIDE (TOX)

TOX is a screening test used to give a general idea of the extent of water contamination by a large number of synthetic organic chemicals. This screening test is sensitive to SOCs containing chlorine and bromine, which accounts for a wide range of possible organic contaminants. These include: THMs; chlorinated and brominated pesticides; organic solvents like TCE and tetrachloroethene; PCBs; and chlorophenols. This is very significant to anyone concerned about their water because it covers about half of the approximately 129 EPA priority pollutants.

For all practical purposes, all organic compounds containing chlorine or bromine are contaminants. TOX then, is **completely** indicative of contamination by these compounds. TOX is probably the single best indicator test for determining water pollution by SOCs. No single test can give a complete picture of water quality, but a negative TOX result will rule out many possible contaminants.

Water can contain organic halides that are not accounted for by other methods. Of all the compounds known to exist in chlorinated drinking water, only one quarter can be measured by conventional testing methods.[1] TOX, however, is the most comprehensive test available for chlorinated and brominated chemicals in water. Many of the heavier, nonvolatile chemicals can be detected **only** by TOX. For many others, TOX is the only easy and convenient method. The TOX procedure does not discriminate against unknown or unclassified pollutants. If there is an SOC containing chlorine or bromine in a water sample it will be detected by the TOX method.

As an example, TOX measurement of a water sample containing 50 ppb chloroform will, of course, bring a result of approximately 50 ppb. However, chloroform will not be identified as the culprit. It will be certain that a chlorinated or brominated organic chemical (or chemicals) is in the water. From this one can decide whether to test further to determine if additional chemicals are present.

TOX is generally most useful to well owners in determining the extent of contamination in drinking water. In addition, the great majority of groundwater pollutants have been found to be halogenated organics. Measuring TOX in well water will identify contaminants of industrial sources. The water is usually not chlorinated, so interference is not a problem. However, a high amount of inorganic halogens like chloride from saline or brackish waters, could interfere with the TOX test.[2] Over-estimation of TOX is possible if the well is relatively deep, and intrusion from lower brackish aquifers is occurring.

Those testing chlorinated municipal water will likely end up measuring disinfection byproducts like Trihalomethanes (THMs). While the water could contain both chlorination byproducts and industrial pollutants, when testing chlorinated water for TOX there is no way to differentiate between the two. Municipal water users may find only THMs and other disinfectant residual organics.

The **detection limit,** the lowest level that can be detected in any sample, for TOX is about 5 ppb. This is not as low as conventional tests which measure specific organic chemicals. However, TOX is sensitive enough to detect pollutants at levels commonly found in water. Anything above 5 ppb, then, can be detected and measured.

The volatile portion of TOX, called purgeable organic chloride (POX) is a separate test, measuring only volatile organic halides in water. Some see POX as a good addition to TOX analysis. The main problem with POX is only about half of the labs doing TOX also routinely do POX.[3] So, it may be difficult to find a lab doing it or it may be expensive.

POX may be decreased in sampling, shipment to the lab, storage, and during TOX analysis.[4] However, if air exposure and turbulence in pouring are kept to a minimum in sample handling in the lab, volatile losses from TOX analysis will not be significant. All one wants to know in the initial tests is whether the water is free of halogenated organics. TOX will indicate this if performed properly. If a positive result is given, other tests can be run to determine exactly what chemicals are present. In that case, a Volatile Organics Analysis (VOA) scan (see chapter 12) should be considered. Then the VOA results can be compared with the indicator results to determine if all pollutants have been detected.

The best advice would be to find a lab that can do TOX, but be aware of the possibility of the loss of volatiles when receiving the results. The EPA requires duplicate and sometimes quadruplicate (samples analyzed twice or four times respectively) analysis of TOX in industrial and regulatory testing. This requirement does not apply to individuals, but duplicate testing is highly recommended. The two results received should be similar.

This is a good quality control measure and will help ensure accurate analysis. Lab workers are more likely to recognize a problem and correct it if two tests are done and measureably different results are obtained.

TOX Sampling & Testing

How water is sampled for TOX is important so loss of volatiles is kept to a minimum between sampling and the time the test is run. The sample container will be similar to the one for VOA, but a larger sample is required, at least 200 ml. As with VOA sampling, the bottle must have a teflon-lined cap and the bottle must be filled completely to eliminate all air bubbles. A bottle can be obtained from the lab that will be doing the test. Samples should be analyzed within two weeks.

Stored chlorinated TOX samples are not stable. If the water is chlorinated, it is likely that it contains residual chlorine which can further react with natural organics after the sample is taken. This means that from the time the water is sampled until it is anlayzed in the lab, including storage or shelf time in the lab prior to analysis, chlorinated organics may be forming. This could give a result higher than the true TOX value.[5] Sulfuric acid and sodium sulfate or sodium sulfite supplied by the lab as a preservative, along with refrigeration, will stabilize the sample and prevent additional TOX formation during transport to the lab and storage time before it is analyzed.[6] Without a proper preservative, the sample should be tested promptly.

TOTAL ORGANIC CARBON (TOC)

Prior to the advent of TOX, Total Organic Carbon (TOC) was considered the best way to determine the extent of organic pollution in water. One drawback of TOC is that it measures not only the manmade organics, but also natural harmless organics common to waterways. The results are a total of all organics present regardless of type or toxicity. TOC does not give the halogenated organic content of the water as TOX does. But, TOC should not be discounted. Not all organic pollutants are halogenated, and TOC will detect these additional organics. TOC then, can complement TOX.

Another important drawback of TOC is that although newer instrumentation will push the detection limit lower, TOC is not very sensitive in detecting low concentrations of pollutants. Currently, the detection limit for TOC is only between 0.1 and 1 ppm. The problem is that compounds could easily exist at toxic levels below the detection limit.

TOC is actually a measure of nonvolatile organic contamination. In the normal TOC procedure, a gas is forced or bubbled into the sample to remove the inorganic carbon (carbon dioxide) prior to actual measurement of TOC. This is a required and necessary step, but volatile organic chemicals are carried out along with the carbon dioxide. Consequently, TOC can only be considered a measure of the **nonpurgeable organic carbon** (NPOC).[7]

Purgeable organic carbon (POC) is a separate test measuring exclusively the more volatile fraction of TOC. Measurement of POC significantly enhances TOC regardless of how it is measured. Volatiles can comprise anywhere between 9% and 50% of the TOC in groundwater.[8] POC fortunately has a low detection limit, 10 ppb. This allows volatiles to be picked up at concentrations near their toxic levels.

One problem is that POC, like POX, is not a common test in many laboratories. It may be difficult to locate a lab that will do it. A lab is much more likley to have a TOC analyzer than to be set up for POC. However, as was the case with TOC and TOX, POC and POX are becoming increasingly accepted and requested. So, it is likely that more labs will expand to include these tests in the near future. VOA, as described in Chapter 12, is another purgeables method which could be utilized. A VOA scan is a good substitute for POC and POX, but the scan must be comprehensive and encompass a large number of chemicals.

TOC can also be an indicator of microbiological contamination. Some use TOC as a low cost test to assure biologically safe water. Groundwater levels should be well below 5 ppm. Any significant change in the TOC level over time could be indicative of pollution.

OTHER INDICATORS

Other indicators discussed in previous chapters are: total coliform, turbidity, HPC bacteria, gross alpha and beta activity, and nitrate. Total coliform, HPC, and turbidity are tests indicative of general microbiological quality. Nitrate pollution, while a concern to infants can also be an indicator of sewage or septic tank contamination of groundwater around drinking water wells. Gross alpha and gross beta activity screening tests measure certain natural and manmade radioactive elements. Table 15-1 lists a number of indicator tests, their advantages, and disadvantages.

The pH of water is also an indicator. It is a simple and inexpensive test used by industry to monitor changes in characteristics of wastewater effluents. The pH of water is a measure of how acidic or alkaline it is. A further explanation of pH is given in chapter 1.

Table 15-1
Common Indicator or Screening Tests

Screening Test	Advantages	Disadvantages
TOC	Measures all chemicals containing organic carbon.	Cannot differentiate between natural organics & manmade toxics; detection limit is not low enough to detect many toxic organics.
TOX	Measures pollutants containing chlorine & bromine, thus, a large number of common & priority pollutants; the test is completely indicative of pollution.	Does not give levels of specific chemicals.
POX	Measures halogenated volatile organics which are common groundwater contaminants.	May be difficult to find a lab doing this test and it may be expensive.
POC	Measures all volatile organic chemicals, both halogenated, and nonhalogenated, which have been found in surface & groundwater supplies; has low detection limit compared to TOC.	May be difficult to find a lab doing this test and it may be expensive.
Turbidity	Good indicator of water treatment efficiency.	Provides little information to homeowners, particularly well owners.
Total Coliform	Considered best single indicator test of overall bacterial water quality.	Is not a direct measure of pathogenic bacteria.
HPC or SPC Bacteria	A direct measure of bacteria in the water sample.	Cannot differentiate between harmful and harmless bacteria.
Gross Alpha Activity	Measures all alpha particle activity - generally indicative of natural radioactivity.	May not be able to determine the presence of Radium-228.
Gross Beta Activity	Measures all beta particle activity - generally indicative of manmade radioactivity.	
Nitrate	Accurate measure of nitrate-nitrogen (highly advisable if you have an infant) & indicator of possible sewage & septic contamination.	Some test interference is possible, reducing the test's sensitivity.
pH	Measures how acidic the water is (should be around 7).	Very limited in providing information; heavy contamination is possible before a noticeable pH change is seen.

16. TESTING YOUR WATER:
Choosing a Laboratory & Testing Methods

"Much water goeth by the mill that the miller knoweth not of."
— John Heywood (1546)

Selecting a laboratory to perform water analysis and choosing specific testing methods are both important considerations in order to achieve a comprehensive picture of water quality. There are a few different types of laboratories, some with distinct disadvantages. These include:

- **Government labs** are traditionally slow in turn-around time and may only do selected tests. Due to financial constraints the government agency may not be able to afford comprehensive analysis.
- **The chemistry or biology departments** at a nearby college or university may do water testing. However, they may not have the financial backing to offer comprehensive testing services. In addition, they may not have the experience that a commercial lab with years of routine analysis would have.
- **The local Health Department** may test private wells upon request, but may only do one or two minor tests like total coliform and pH. Both are useful, and helpful, but more comprehensive testing is necessary to

determine overall water quality.

* **The party believed responsible for the contamination,** if known, is unlikely to volunteer to test the water. If they do, they are not going to return incriminating results that could be used against them legally.

Lab results can be requested from municipal water utilities. These are a matter of public record. But the chemicals they test for are generally only those which are required by law, and these continue to be minimal. Small systems often cannot afford to test for many of the important chemicals that are a part of a larger system's routine analysis. Some even avoid testing for those required by filing for variances or exemptions. Lab reports will give levels of the chemicals that are regulated, but remember that these are the levels at the plant. A lot can happen between the treatment plant and the tap: THMs can increase, lead can go up, and waterborne bacteria can multiply.

The best way to obtain comprehensive analysis of drinking water is to have a commercial lab do the testing. There are many labs to choose from and, as a paying customer, you have complete control over what testing is done. The yellow pages will outline the types of labs in the area. Look under, **laboratories, testing; water analysis;** and **water testing** for an environmental testing laboratory that analyzes water. Most environmental labs analyze water, but, some deal exclusively in industrial air samplings. The American Council of Independent Laboratories, Inc., 1725 K Street NW, Washington, DC 20006 can provide a list of labs in the area that do water testing. There are also a few labs around the country that specialize in testing for homeowners. They provide sample containers, sampling instructions, and packaging for mailing samples to the lab (see Appendix V).

Most environmental labs are in business because federal regulations require certain companies to have their effluent discharge to waterways monitored by an outside laboratory. The Clean Water Act of 1977 required some commercial companies to obtain a discharge permit. It mandated that they send samples to independent environmental labs on a regular basis to prove their compliance with the law. This brought about somewhat of a boom in laboratories. With the passage of the Safe Drinking Water Act (SDWA) amendments, labs also moved to accommodate the demand by water utilities which needed to analyze for certain drinking water parameters.

Laboratories are expensive operations. An enormous amount of capital is necessary to start a laboratory. An instrument for heavy metals analysis costs at least $40,000. The instruments necessary to do organics analyses can run several hundred thousand dollars. Add to this a host of laboratory glassware, chemical reagents and standards, as well as the technical personnel needed to perform the tests, and it means a sizable startup cost. Due to these high initial and ongoing costs, most labs with comprehensive testing services are either backed financially by a larger corporation, or were in existence prior to federal regulations and the advent of more sophisticated water testing methods developed in the late 1970s. More recently, many labs have expanded or corporations have formed lab sections to obtain EPA contracts to analyze samples from hazardous waste sites. Those that were not awarded the contracts had to step up their list of industrial clients to offset the costs of expansion.

What to Look for in a Commercial Lab

A lab should be certified by the Federal EPA for all the tests requested. Some labs are only certified for metals and/or **"wet methods"** like nitrates, fluoride, and hardness. Certification for the more difficult SOC methods like VOAs and pesticides is harder to obtain, so not as many labs are certified for them. Make sure the lab doing these tests is certified for them. In addition to EPA certification, state certification is a plus.

Winter is often a relatively slow time at labs. Thus, results may be returned faster, and the chemist may have more time to devote to a specific situation or an individual's questions. Regardless of the time of year, do not be in a hurry to receive the results. Most labs offer a two-week turnaround time — time between sample delivery and the return of results. But do not hold them tightly to this. Be nice, they will appreciate it. Laboratory workers are used to clients screaming for results and are constantly under deadline pressure. A smile and kind words will go a long way.

When testing time-dependent parameters like VOAs or total coliform, where holding-time before analysis is a critical factor, look into labs close by. If there is no choice but to mail them, make sure they are put on ice during shipment and are received by the lab within 24 hours. If the testing is being done in response to some kind of local controversy, it might be advisable to choose a lab that is some distance away. The most important criteria for choosing a lab, however, is its quality assurance and quality control program.

Quality Assurance and Quality Control

In any analysis, one or more standards (a known amount of the specific chemical being analyzed) are tested along with the sample. Simply stated, the response of the standard (the amount which is known) is compared with the response of the sample (the amount which is unknown).

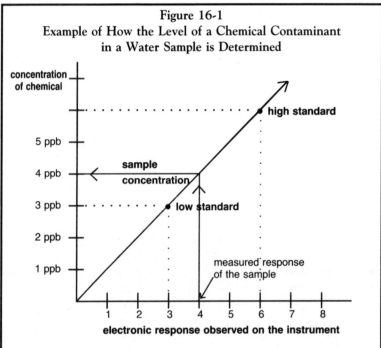

Figure 16-1
Example of How the Level of a Chemical Contaminant in a Water Sample is Determined

In any test, chemical standards (the concentrations which are known) are measured on an instrument to obtain an electronic response. With this information an electronic "graph" can be plotted. When the sample (with an unknown concentration of the chemical) is measured for a response, the concentration can be easily determined.

This comparison determines the amount of the contaminant in the sample (Figure 16-1). Often two standards are run on a particular test, one high level and one low. This is done because precision often diminishes with higher concentrations.[1]

The EPA has approved analytical methods for testing water that must be rigidly followed. There is no magic. Things can go wrong in a lab as easily as anywhere else. Quality Assurance/Quality Control (QA/QC) guards against erroneous test results.

Quality Control (QC) is all the actions taken by the laboratory to insure all factors affecting the laboratory measurements are accounted for. This includes problems within each sample, lab technician performance, and instrument reliability. **Quality Assurance (QA)** includes every action taken, both in and outside the lab, to be sure that the results generated by the lab agree with other labs within a network.[2] QA/QC, then, strives to insure that testing done on the same water sample by different technicians, at different labs, at different times will give similar results.

QA/QC determines the **precision** and **accuracy** of the test being done. These two words, although used interchangeably, have very different meanings when pertaining to lab work. Precision is a measure of how well a group of results agree, or how reproducible the test is. In other words, if the same sample was run five times and the results were close together, it would mean the precision was very good (Figure 16-2). Accuracy is a measurement of how close the resulting number obtained through the analysis is to what the actual amount of the contamination is in the sample. That is, how close the measurement is to the true value (Figure 16-2).

There is, of course, no way of knowing at what level a contaminant is present in the water. That is why it is being tested. But there are ways to know of the accuracy of the test procedure itself. This is done by obtaining a water sample containing a known amount of the chemical being tested, sometimes called a QC check sample. It is tested along with the other samples to make sure the entire test is accurately performed. If the QC check sample is analyzed and the result is accurate (very close to the known concentration of the QC check sample), then the water sample result is considered accurate.

Every lab should have a standard QA/QC policy that covers all samples analyzed. If the lab is of any size, anyone telephoning will probably be turned over to a salesperson. He or she should know the tests the lab can perform, explain the lab's QA/QC policy, and provide a written QA/QC outline. If a lab cannot provide a written QA/QC policy, do not bother to send them samples. Lack of a structured QA/QC policy is a sure sign of poor technique and a lab putting out questionable results. As a rule of thumb, an adequate QA/QC program should take up at least 10% of a lab staff's time.[3] This could pose a significant problem for small labs that may not be able to devote that much time and money to a QA/QC program. It is, nonetheless, necessary and very important, even with the increased costs. This is not to say that smaller labs do not have adequate QA/QC policies. Most do and have other advantages as well. One is more likely to get personal treatment at a smaller lab, and not get lost in the shuffle of a larger one.

A number of different QA/QC measures are used by laboratories. These include:

Figure 16-2
How the Words Accurate and Precise Differ
(bull's eye represents the true amount)

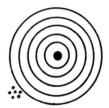

PRECISE	**ACCURATE**
shots are far from the bull's eye, but very close together indicating consistency and repeatability	shots are scattered but very close to the desired objective

- **External** or **Interlaboratory Quality Control Check Samples** are samples of known concentration obtained from an outside source, like the EPA or a state agency. One should be analyzed with each group of samples tested. What the *QC check sample* does is regulate everything happening in the lab — whether the standards were prepared correctly, or whether the instrument is responding properly, etc. Without this outside comparison, the lab would be operating as an island, totally isolated, without any proof that it is generating accurate results. Scientists agree that while external QC is important, it is not a substitute for a good QC program within the lab.[4]

- **A Method Blank** is used to determine if the samples have received any contamination during the testing procedure from lab glassware, the standards, or other sources in the lab. A *method blank* is ordinary laboratory-grade or pure water that goes through the entire analytical procedure along with the samples. In PCB analysis for example, if PCBs are found in the method blank and the samples, it is quite possible that the contamination occurred in the laboratory. Therefore, the PCBs found in the samples may not have been in the water when it was first sampled, and the entire procedure should be repeated. A blank should be analyzed with each group of samples and taken through the entire method procedure just like all samples and standards.

- **A Duplicate** — is the sampling of the source twice, like splitting the sample. This gives two identical samples. Thus, their analysis should give very similar or nearly identical results. This determines whether the sampling was done in such a way that it is representative of the source from which it was taken. An example of **non**representative sampling would be to take a sample immediately after turning on the water faucet, and then taking another after letting the water run for five minutes.

- **A Replicate** — analysis is essentially the analysis of the same sample twice. It evaluates the lab's ability to "prep" the two samples identically and do a precise analysis. In duplicate/replicate analysis, as with lab analysis in general, the results will not be *exactly* identical. A minimal amount of variability is to be expected.

- **Spikes** — are used to determine what effect the sample makeup or matrix has on the analysis. Water may contain solids, minerals, or other chemicals that hinder the actual anlaysis. Suppose a TOC sample result is *below* the detection limit of 1 mg/L. To prove that there are no interfering components in the sample which are masking the organic carbon, the lab technician may "spike" the sample with some organic carbon, say 10 ppm. If the spiked sample is run and the result is 9.5 ppm, then the recovery is 95% (good recovery) which indicates no internal interference within the sample.

- **Blinds** — Obviously when external or internal QC samples are run by a lab staff, they are given preferential care and handling in their analysis. Consequently, the best possible performance that a lab can give is devoted to the QC samples, as opposed to routine samples. To determine if bias is occurring, blind samples are slipped in to the sample flow in the lab. Blind samples are those of known concentration that are given fictitious client names and submitted by lab management along with other incoming samples. This is done to test the impartiality of the lab technical staff.

The results of the quality control that was done with the samples submitted can be requested. They may not actually have been done **on** those samples, but the QC done on samples in that group would be helpful in interpreting the precision and accuracy of the results. At least one QC sample should be run every ten to twenty samples. Make sure the lab follows this rule.

A copy of the results of a recent **Laboratory Performance Evaluation** or **Laboratory Audit** that the lab participated in can be requested from the laboratory. These evaluations are conducted by the EPA to determine the accuracy of the laboratory's testing. The report will list the true value and the lab's result for a number of tests.

Detection Limits

The *detection limit* of a particular test is the lowest concentration of the chemical(s) that the analytical procedure can reliably detect. This will vary from test to test and from chemical to chemical. The detection limits for common SOCs appear in Appendix III. This limit may be a function of the procedure itself or the instrument used in the test. The detection limit is, however, generally a function of the entire analytical procedure and not just the limitations of the instrument. The detection limit for a particular procedure may also vary between labs. The detection limit for TOX for example, is generally considered to be 5 ppb[5] but some labs report 3 ppb, while for others 10 ppb is their detection limit.

Detection limits could vary on different samples as well. Widely different samples from a variety of locations contain equally variable interfering components, from suspended solids to chemically reactive natural inorganics. These interfering substances can increase the detection limit, that is, reduce the sensitivity to one or more specific chemicals being tested. Interference and loss of sensitivity to contaminants is usually associated with wastewaters or hazardous waste where high levels of interfering substances are common. In the case of drinking water samples though, components should not exist that would significantly reduce sensitivity and raise the detection limit.

Sample Splitting

Splitting the sample can be a useful psychological tool depending on who is analyzing or sampling the water. If there is a serious contamination problem and a government agency or the industry believed to be causing the pollution is in charge of the testing, anyone involved would want assurance that the testing is done correctly. Requesting a separate sample at the time of sampling will put them on their guard to ensure accurate analysis. Even if the decision is made not to have the samples analyzed by another lab, the belief that the results will be compared with others may elicit more careful treatment. Be there at the time of sampling and request an exact duplicate sample or set of samples. Make sure the sampling methods and containers are identical, and the sampling technique is the same.

Sampling for Water Analysis

When sampling, you should know what special considerations are necessary for each test requested. For example, some organics test bottles must be pre-rinsed with the solvents that will be used in the extraction.

VOAs must be sampled in special bottles to assure that the volatiles do not escape. Glass bottles with plastic teflon-lined caps must be filled completely to eliminate air bubbles (see Chapter 12). It is best to obtain sample collecting bottles from the lab. They should be happy to do this, in fact many labs may require it. They should also supply specific sampling instructions. If a lab cannot or will not provide sampling containers or sampling instructions, find a lab that will.

As discussed in previous chapters, tests for some chemicals require that a preservative be added to the sample at the time the sample is taken. Also, a time limit or **holding-time** is required for many tests. This means that the sample must be analyzed within a certain time period. The preservatives and holding times for some common tests are given in Table 16-1. Be sure to mark the date and time the sample was taken so lab personnel are aware of any danger of exceeding the holding-time.

Interpreting The Results

All labs will report the results of their analysis in writing. The lab report will look something like Figure 16-3, listing the test and the corresponding result or concentration. It is easy to interpret these results. A test result can only be reported in one of three ways.

1. The results of an analysis will be given as "less than" the detection limit if nothing is found. For example, for TOX if nothing is found, the result will be expressed as $<$ 5ppb ($<$ is the symbol for "less than") because 5ppb is considered the detection limit for TOX.

2. If the test is positive, the result would be written as the amount of the contaminant found, i.e., 25 ppb (ug/L).

3. If the result is something like $>$ (greater than) 50 ppb, there is a problem. The lab is incapable of reaching the normal detection limit due to serious interfering substances in the water and you should investigate why. Do not be intimidated by chemical abbreviations or symbols if they are used. Simply ask the lab to write out any chemical names.

So what do you do with a positive lab result? What if the water is contaminated? A mistake many people make when they find a pollutant in their well, is to assume that they have found all contaminants. It is very rare to find a single pollutant, particularly an SOC, in a groundwater supply. If one is discovered, usually further testing will reveal more. The initial finding of water contamination by a single chemical may be the first plume in a line of overlapping plumes of contaminants that are moving through the aquifer from the source toward the well. Since different chem-

TABLE 16-1
Recommended Volumes, Containers, Preservatives, and Holding Times for Water Analyses

Chemical Analyses	Vol. Req. (ml)	Container (Plastic Glass)	Preservation	Holding Time
Nitrate	100	P.G.	sulfuric acid, & Cool, 4C	48 hrs.
pH	50	P.G.	analyze immediately	2 hrs.
Turbidity	100	P.G.	Cool, 4C	48 hrs.
Metals	100	P.G.	nitric acid	6 mos.
Aluminum, Antimony, Barium, Berylium, Cadmium, Calcium, Chromium, Cobalt, Copper, Iron, Lead, Lithium, Magnesium, Manganese, Molybdenum, Nickel, Potassium, Silver, Sodium, Strontium, Thalium, Tin, Titanium, Vanadium, Zinc				
Chromium VI	100	P.G.	Cool, 4C	24 hrs.
Mercury (Cold Vapor)	100	P.G.	nitric acid	28 days
Fluoride (Specific Ion Electrode)	300	P	none required	28 days
MICROBIOLOGICAL ANALYSES				
Total Coliform	100	P.G.	Cool, 4C*	6 hrs.
Standard Plate Count	100	P.G.	Cool, 4C*	30 hrs.
ORGANIC CHEMICAL ANALYSES				
Purgeable Organics (VOAs)	2 x 40 ml	G, Teflon Septum	Cool, 4C*	14 days
Non-Volatile Organics	2 liters	G, Teflon Capped	Cool, 4C	7 days until extraction 30 days after
Phthalates	2 liters	G, teflon cap	Cool 4C	7 days until extraction, 30 days after
PAHs	2 liters	G, teflon	Cool 4C	7 days until extraction, 30 days after

(*Thiosulfate required if sample is chlorinated.)

RADIONUCLIDES†				
Gross alpha activity	2 liters	P.G.	nitric acid	6 mos.
Gross beta activity	2 liters	P.G.	nitric acid	6 mos.

(†Other radionuclides require specific preservatives, check with the lab.)

ORGANIC INDICATOR METHODS				
Total Organic Carbon (TOC)	100 ml	G	sulfuric acid	7-10 days
Total Organic Halide (TOX)	500 ml	G	none required	15 days

Source: U.S. EPA

Figure 16-3
Sample Lab Results for a Well Water Analysis

XYZ Labs
13000 Constitution Avenue
Middletown, ID 00000   Client: John Q. Public

Parameter	Concentration
TOC	< 1 mg/L*
TOX	50 ug/L†
Total Coliform	< 1 coliform/100 ml
Nitrate	< 1 mg/L
Metals	
Pb (lead)	
1 (faucet)	50 ug/L
2 (interior plumbing)	30 ug/L
3 (well)	< 0.1 ug/L

Interpretation of Lab Report

- TOX, well above the detection limit of 5 ppb (ug/L) indicates contamination by halogenated organics. Further testing is necessary, probably a VOA scan.

- Nitrate and TOC are less than the detection limit.

- The total coliform result indicates a lack of bacterial contamination.

- The lead results indicate that flushing the in-house plumbing completely before use may be all that is needed to reduce the elevated lead levels in this case. However, to play if safe, this homeowner should retest for lead and/or consider some treatment to reduce water corrosivity.

*mg/L = ppm
†ug/L = ppb (see Chapter 1)

icals move at different speeds in comparison to the water flow in the aquifer, this advance of consecutive plumes would result in new contaminants showing up in the water at different times. This scenario could continue for decades even if the initial source is cleaned up. This can occur in the case of a landfill or other source of a wide variety of different chemicals.

When a receiving positive test result do not accept that level as being carved in stone. Test again. Due to the nature and unpredictability of groundwater, past experiences with contamination show that where several consecutive samples were tested, different results were obtained. In Uncas-ville, Connecticut, for example, one well contaminated with tet-rachloroethylene tested at 17 ppb, 236 ppb, and 147 ppb on different occasions.[6] While this may be a sampling error, it may also be due to the movement of tetrachloroethylene within the aquifer. The lowest concen-trations of a pollutant will be at the edges of the plume. In this case, then, exposure levels are variable.

The next big question when you look at a positive lab result is, what is a safe level? Level is everything and level is nothing. The concentration of a contaminant is very important in terms of health effects, but Maximum Contaminant Levels (MCLs) do not always reflect this. In some cases, MCLs are continually changing, so what is legally safe one day may not be the next. The basis for any regulatory level is determined through studies and research that can be incomplete, conflicting, and inconclusive.

If more than one compound is discovered, it would be even more difficult to predict the *synergistic* effects. No toxicologist can project the health effects of the possible interaction between more than one chemical. Man lacks full understanding of environmental avenues of chemical expo-sure, chemical movement in the human body, and how exposure manifests itself as disease. Uncertainties abound when trying to match environmental pollutant exposure with health effects. How chemicals may affect humans will depend on the length and level of exposure, and the specific chemi-cal(s) involved.

Any detectable level of a manmade pollutant is cause for concern. One could wrestle with the scientific literature and obtain as much infor-mation on the chemical as possible to figure out what a safe level is. But anyone undertaking this will likely still be left with plenty of questions. Consulting a professional on the significance of the levels of contaminants in the water is an option, but he will likely quote the MCLs as being safe levels or compare the level to other sources of chemical exposure.

Knowing the MCL or Maximum Contaminant Level Goal (MCLG) for the chemical present will give you an idea of the severity of the contamination. For inorganics, like metals, safe levels are better understood because many of them exist in natural water systems where the water contacts the minerals containing them. However, most MCLGs for SOCs are zero. So, if the water has a detectable level of an SOC, it would be highly advisable to have some kind of treatment as well as further testing to determine what else might be present. In cases of extreme contamina-tion, alternate water sources may have to be explored. Needless to say it would be wise to determine the source of the pollution and, if possible, eliminate it.

According to the National Cancer Institute, "The idea that there is a safe dose of chemicals may be conceptually valid, but safety cannot be established by any experimental method now available."[7] With little alternative, regulators act based on what information is available. But is this enough? Do safe levels in fact exist? Samuel S. Epstein, Professor of Occupational and Environmental Medicine at the University of Illinois, states, "There is no safe level of exposure to a carcinogen."[8] According to Dr. Irving J. Selikoff of Mt. Sinai Medical Center, low level exposure should not be underestimated.[9] Even the EPA's viewpoint is that low levels of organic chemicals are a significant risk to health and require regulation.

In short, no one knows what safe levels are for contaminants. There are no set rules to follow or a concrete universally applied method for assessing risk of chemical exposure and determining a safe level. Estimates and educated guesses abound, but all that has really been determined is which chemicals are more dangerous than others. The seemingly endless number of variables, and the uncertainty of scientific data make each situation different. A personal assessment of risk, however cautious, is as good as any. Determining a level where risk is acceptable is not a decision for scientists. It is a decision that should be left to society, its legistlative representatives, and ultimately the individual. In the end, you must ultimately decide what level is safe and what is not.

Cost of Testing

The cost of different tests can vary greatly between labs, so it may pay to shop around. The cost of testing water will depend on:

1. **The tests selected.** Pick the ones that will supply the most information or cover the greatest number of possible contaminants, like indicator tests and VOA screens. Request tests by method, do not request analysis of individual chemicals. The customer can generally get more for their money, because all the chemicals in a method are often tested anyway. Screening methods like TOC and TOX have the advantage of being comparatively inexpensive. High-tech equipment and highly skilled personnel are not required for these tests.

2. **The lab.** Overhead and pricing structures can vary with labs as with any other business.

3. **The QA/QC program.** The stronger and more rigid the QA/QC policy, probably the higher the cost of tests to offset the time devoted to it. Still, it is imperative that the lab has a QA/QC policy and implements it in the analysis of its samples. Unusually low prices may be the sign of poor or nonexistent QA/QC.

TABLE 16-2
Approximate Prices for Common Tests

Tests	Approx. Price Range $	Test	Approx. Price Range $
pH	$3–7	**SOCs**	
hardness	8–25	phenols (EPA 604)	$80–160
fluoride	12–30	phthalates (EPA 606)	160–228
Nitrate - Nitrogen	12–40	PAHs (EPA 610)	150–228
Nitrate - total	18–45	GC/MS volatiles (EPA 624 or 524)	150–300
Metals	9–20 each	GC/MS acids (EPA 625) - phenols	125–300
(may be less per metal if more than one tested)		GC/MS base neutral (EPA 625) - phthalates, PAHs etc.	200–385
Mercury - Cold Vapor (special test)	30	GC/MS base neutral & acids (EPA 624 & 625)	292–500
Chromium VI	16–25	GC/MS pesticides	200
(special test aside from chromium total)		EPA 601 or 502 (volatile halogenated organics)	50–164
Microbiological Methods		EPA 602 or 503 (volatile aromatic hydrocarbons)	50–150
turbidity	7–25	THMs	30–75
total coliform	13–40	EPA 608 (pesticides & PCBs)	80–225
SPC	12–25	SDWA pesticides and herbicides	150–200
Radionuclides		Herbicides (2,4-D; 2,4,5-T)	80–160
gross alpha activity	40–65		
gross beta activity	40–60	**Other Tests**	
gross alpha and beta activity	30–100	asbestos	35–90
radium 226	40–128	TOC	25–55
radium 228	75–189	TOX	45–90
radium total (226 and 228)	100–200		
radon	70	**New Age Pesticides (see Appendix IV)**	
uranium	20–70	EPA 504 ... 50 EPA 508	120
strontium 90	222	EPA 505 ... 120 EPA 515	120
		EPA 507 ... 120 EPA 531	120

4. **Whether the customer or the lab collects the sample.** You can save considerable cost by sampling and delivering the sample to the lab yourself. However, if litigation is anticipated, have a commercial lab do **everything.**

Table 16-2 lists the approximate prices you might expect to pay for generally available tests based on a random survey of laboratories. These prices may vary depending on geographic area and differences in the labs themselves. However, be wary of prices too far out of line from this list. Some labs will send out samples for tests they cannot do in-house to other labs for analysis. For example, tests for radionculides are not common to many labs and samples are often sent out to another lab specializing in them. Also, smaller laboratories often send out SOC tests. This is a common practice, and adding a small handling charge for sending out samples is acceptable. But beware, an unscrupulous lab will mark up the cost as much as 400%. The lab may even list the test on their laboratory price list, giving the impression that it is done in-house.

While you might be concerned about the expense of water testing (and rightly so — some of these tests are expensive), ignoring the situation could be even more expensive. It would be foolish to spend up to $1,200 plus maintenance costs on an in-home point-of-use treatment device if it is not needed. The money would be better spent on comprehensive testing to determine if treatment is necessary at all.

It would also be unwise to buy an expensive house or have one built without knowing the quality of the water, particularly if there is only one source. It might be possible to drill a new well, but that could be as much as $10,000 and there is no guarantee the new well will not be contaminated. You certainly would not purchase a house without having it thoroughly inspected by a certified inspector to make sure it is structurally sound. In the same way, assuring the water supply is uncontaminated is the safeguard of a large investment by spending considerably less. If you are having a house built or are considering building a house, particularly in the country, the time to test is when the well is drilled, before the house is built. Finding contaminated water could significantly alter the plan. Today, water testing should also be a part of routine home maintenance.

The decision of what to test for is ultimately up to you. All situations and concerns are different and unique. Living near a chemical plant, waste site, or other possible source of a wide variety of contaminants, will probably mean a comprehensive testing strategy, along with some detective work as well. Avoid getting caught up in media-panic or the "toxic chemical-of-the-day club." Evaluate the situation and approach any testing strategy logically.

17. TREATING YOUR WATER:
Home Treatment Methods and Devices

"In rivers, the water that you touch is the last of what has passed and the first of that which comes: so with time present."
— Leonardo DaVinci (1452–1519)

Many people, concerned about the safety of their water, run out and buy a water filter, attach it to their faucet, and sit back erroneously thinking their water is now free of all possible contamination. An advertisment may list all the tremendous things a new device can do but you may not have the water quality problems they so artistically outline. Fancy words can be misleading. Many units can remove only taste and odor, but manufacturers make claims about the reduction of synthetic organics, however minimal that may be.

In order to fight the enemy, you must know the enemy. So, it is important to first test the water to determine if any contaminants are present. Based on the test results, the decision can be made as to whether a home water treatment device is needed, and if so, which type. If there are no problems, treatment would be a waste of time and money. In fact, if the water quality is good, a home treatment device may degrade the water by introducing bacteria without any reciprocating benefit.

Home water treatment units, or *point-of-use* **(POU)** devices as they are called by the water treatment industry, are classified by the groups of chemicals that they can remove or neutralize. There are five types: basic filters, which only remove suspended solids or taste and odor, not pertinent contaminants; activated carbon adsorption filters; reverse osmosis units; distillation systems; and disinfection units. Some devices use a combination of these in one unit. As might be expected, combination units will remove a wider variety of contaminants. But, no single treatment unit offers complete removal of all possible contaminants.

Most treatment units remove only a percentage of the contaminants that they are designed to remove. POU devices are a salvation and blessing to consumers in many respects, but technology is not so far advanced that manufacturers can guarantee that absolutely no amount of a contaminant will be found in the treated water. On the positive side, many well manufactured devices can remove well over 99% of the contaminants that they are designed to reduce.

It is estimated that less than 1% of the water coming into the house is actually used for cooking and drinking. Unless there is a serious contamination problem it would only be necessary to treat this water. It will place considerably less strain on the treatment system, and would also save money due to less frequent replacement of filters. Based on this, a point-of-use treatment unit near one faucet may be a good idea. However, in cases of high volatile organic contamination or elevated radon levels where contaminants may dissipate into the air from showers, toilets, or dishwashing, it may be advisable to treat all water entering the house with a *point-of-entry* **(POE)** device. Compounds released to the air can be a hazard to those inhaling them. Treatment for these airborne contaminants may not be complete with a POU device.

Activated Carbon Adsorption

Activated carbon removes halogenated organic compounds (those that contain chlorine or bromine) as well as some other organic pollutants. Activated carbon is carbon from wood, coal or lignite that has been ground up and *activated* — heated at a controlled temperature and pressure to promote active sites where pollutants can be adsorbed. *Adsorption* is a physical process that simply means certain water pollutants are more attracted to the surfaces of the carbon than to the water. Therefore, they can be pulled out of the water solution.

Granular Activated Carbon **(GAC)** home treatment units are the most common. Large, heavy compounds with a low solubility in water, like halogenated organics, are most effectively removed by activated carbon. However, other nonhalogenated compounds, like benzene, are also removed to some extent. This does not mean GAC is the cure-all. Well manufactured GAC devices are the most efficient at halogenated organic chemical removal. However, they are inferior to Reverse Osmosis in particulate, inorganic chemical, and heavy metal removal.

What many people who own an activated carbon unit do not realize is that the filter must be changed regularly to avoid leakage. The longer the filter remains in use the less attraction the carbon has for the more weakly adsorbed chemical compounds. It may get to the point where all the active sites are used up. When this occurs, the organics will come off the filter, resulting in a higher level of organics in the treated water than in the water going into the filter. A worn-out filter is worse than having no filter at all. How often a filter must be changed is dependent on the amount of carbon in the filter, the degree of usage, the extent of water contamination, and the filter's specific design. The higher the concentration of pollutants in the water, the sooner the active sites will be used up and the filter will loose its effectiveness. Use manufacturer's suggestions for filter changes as a guide, but change it more frequently than suggested. The EPA suggests three months as a general rule, but says deviations from this can go both ways.[1]

There are many types of GAC filters on the market with varying degrees of effectiveness. How well a specific filter operates depends on:

- the quality of the water;
- the design of the filter;
- the amount of activated carbon in the filter;
- the quality of the activated carbon;
- the amount of water passing through the filter; and
- the time the water spends in contact with the activated carbon.

The more activated carbon in the filter, generally the more effective the device. Tests on different brands of carbon filters, including those coupled with R-O units, determined contaminant reduction efficiencies ranging from 76% for small faucet-mounted units to 99% for some larger capacity line-bypass units.[2] The type of activated carbon used is also a factor in efficiency. GAC has been proven to be superior to powdered carbon and is most often used in filter units.

Manufacturer's design is also a critical component of an effective filter. Though reduction of halogenated organics is usually greater for units which contain more carbon, design proved more effective than the amount of GAC in at least one test. The study compared two systems, each with a 2500 gallon capacity. One with 527 grams of GAC showed an average of 61% THM reduction. The other contained 1120 grams of GAC — more than twice the amount — and had an average THM reduction of only 43%.[3] Filter design, then, can be a considerable factor in filter performance.

Some filters force the water to take the long route through the carbon, allowing it to spend the most time in the carbon. But, other filters allow the water to take short cuts, thereby reducing the contact time. The filter design should be such that the water is forced to run the entire length of the filter. This ensures a longer contact time between the GAC and the water, leading to more efficient removal.

The flow rate of the water running through the filter will also determine how well the water is treated. A rapid or gushing flow will not keep the water in contact with the carbon long enough to sufficiently remove the contaminants. Trickling is the optimal rate, allowing the most contact time between the filter and water. Some units, however, can handle slightly faster flow rates without compromising removal efficiency.

A few water constituents can reduce removal efficiency. A high Total Organic Carbon (TOC) content in the water can decrease the ability of the GAC to absorb VOCs.[4] Some GAC units come with a sediment pre-filter to remove suspended solids that can reduce the GAC's effectiveness. A prefilter can be purchased separately for those that do not. They run under $50, and replacement filters cost less than $10.

Activated carbon is made from several natural organic sources. Lignite carbon usually has a greater pore diameter which makes it ideal for removal of large molecules. Activated carbons from bituminous coal removes a broader range of organic pollutants, including the larger molecules. Coal based GAC has proven to give the best overall results in terms of removal efficiency for the longest period of use.[5] A filter with a large amount of lignite carbon may not be as effective as a filter with a smaller amount of bituminous carbon.

The mutagenic activity of water, an indicator of genetic human hazards, is greatly reduced by GAC filters. Organic contaminants contributing to a water's mutagenic activity are numerous and, as discussed in Chapter 15, cannot be quantified by any current, conventional methods. GAC has been shown to completely remove mutagenic activity in initial treatment, and consistently remove more than 87% after three months of operation of the filter.[6] In addition, radon is also effectively removed from water by GAC units.[7] GAC filters would also be ideal for those on municipal water systems who want to reduce THM concentrations in their water.

If you elect to purchase a GAC unit, you should remember to:

- change the filter well before the manufacturer's suggested date;
- use the slowest flow rate possible;
- flush out the filter before using the water; and
- retest the water after installation to be sure filter is operating correctly.

Types of Home GAC Filters

There are four basic types of GAC units available: faucet-mounted; line-bypass; in-line stationary; and portable pour-through (Figure 17-1). Only a general comparison can be made between the smaller, less expensive faucet-mounted models and the larger, more costly undersink units. Most of the small faucet-mounted activated carbon devices are designed to control only the taste and odor of the water with no claims of removing SOCs.[8] Generally, they are too small, the design is inadequate, and the flow of the water out of the faucet is too fast to significantly remove organics.[9] The large capacity undersink units are considered the most efficient due to a larger amount of activated carbon and longer contact time between the carbon and the water. In a study of filters by Gulf Southern Research Institute for the EPA, scientists found that chemical reduction efficiency ranged from 76% for faucet units to 99% for line-bypass (under sink) units.[10]

A quality GAC filter should remove 90–100% of THMs and other chlorinated organics. Some well designed line-bypass counter-top units that operate by a hose connection to the faucet are also very effective. Portable pour-through filters can be effective, but they must have an adequate amount of carbon. Generally speaking, a contact time of over 35 seconds and a GAC weight of 1.5 pounds gives the greatest THM reduction.[11]

There are GAC units under $100, but some run as high as $500. Price is not always an indicator of efficiency, so shop around. Also, be advised that replacement cartridges could run from under $10 to as much as $100.[12]

Bacteria Growth in GAC Units

In municipal water supplies, a certain amount of residual bacteria remains in the distribution system. It is not all killed by the chlorine.

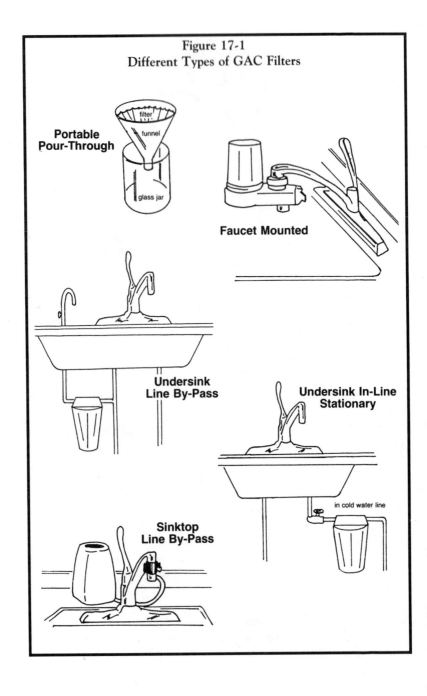

Figure 17-1
Different Types of GAC Filters

Pathogenic bacteria can exist in household plumbing in numbers too low to cause disease, but when the bacteria reaches the moist stagnant environ-ment of the GAC filter it could multiply to levels high enough to be harmful. Even a small amount of bacteria trapped in the filter can colonize it and grow to significant numbers. In addition, activated carbon units, while removing halogenated organics, also remove the inorganic chlorine and trap natural organics and sulfur. The loss of bacteria-killing chlorine and the addition of these nutrients make GAC units ideal for breeding bacteria.[13] Some studies reported that due to colonization and growth of bacteria in the filter, more microoganisms were found in the water coming out of the filters than in the untreated water.[14]

Numerous studies have found high bacterial contamination in GAC filters after periods of non-use. Tests by Gulf Southern Research Institute found the highest populations of bacteria in the first water out of the filter each day.[15] When water samples for bacterial analysis are taken, even without a filter in use, the results are usually lower when the sample is taken after letting the water run for a while.[16] Not allowing lines to flush after non-use can cause significantly elevated bacteria levels.[17]

Humans as well as other animals can tolerate or adapt to moderate levels of bacteria in water. However, the health effects of significant levels of a wide variety of bacteria and microorganisms are not clear. Some view the use of POU activated carbon filters without accompanying disinfection as a potential hazard to health.[18] Consequently, with a lack of clear knowl-edge about the effects of biological contamination, it would be advisable to know the biological quality of the water before using a GAC filter.

Concern about the ingestion of bacteria and opportunistic pathogens led the Canadian government to consider a ban of POU carbon filters. Instead, due to the absence of known illnesses linked to carbon filters, they resolved to caution against their use where the microbiological quality of the water is unsafe or unknown.[19] Scientists do not recommend the usage of POU devices in cases where the coliform standard is not met.[20]

Confined undersink GAC units are regularly subjected to heat or increased temperature from kitchen appliances and hot water from the sink. This kind of environment gives microbial growth a big boost. Insu-lation of the unit to prevent exposure may be a good idea. Cellulose paper filters have also been implicated as a source of microbiological problems due to the breakdown of the filter during usage.[21] It is probably best to avoid these when purchasing a unit or replacement filters.

Individual wells should have no bacterial contamination at all.[22] If a coliform test reveals that there is some bacterial contamination, it should be easier to remove it in a small system than it is for a water company to control it in a large distribution system. The best action is to remove the source of the contamination. If this is not possible, just as it may not be posible to remove the halogenated organic source that is the reason for the filter in the first place, then disinfection methods may be in order.

Densities of bacteria found in point-of-use devices vary due to: filter design; water temperature; bacterial levels in the influent water; flow rate; types of bacteria; efficiency of halogen removal; carbon surface area; and the length of time the filter cartridge is used.[23] Once colonized, a device will discharge the bacteria indefinitely. Precautionary measures like flushing the filter before drawing water are often the only consistent preventive action considering the randomness of biological contamination and the inability to have samples rapidly analyzed.

To guard against biological contamination of GAC and other home treatment devices, follow these guidelines:[24]
- only use the unit for treatment of microbiologically safe water;
- flush the unit after periods of non-use — the longer the period without usage, the longer the flush;
- change the filter according to the manufacturer's specifications, or even more often; and
- follow all manufacturer's maintenance instructions.

Reverse Osmosis

Reverse osmosis (RO) is a comparatively new mode of treatment. Its operation utilizes the normal pressure created in water lines to force the water through a semi-permeable or selective membrane which separates contaminants from the water (Figure 17-2). Treated water emerges at the far side of the membrane, the outlet, and dissolved solids and impurities are left behind.

RO units are designed to remove 90–99% of most dissolved pollutants as well as many kinds of bacteria.[25] A study of pesticide treatment by reverse osmosis reported a 98–100% removal of chlorinated and organophosphorus pesticides. Other pesticides, like atrazine, were found to be less effectively removed.[26] Tests on pesticide removal by RO in a 1983 study found that 90% of the endrin and methoxychlor, but only 40% of the lindane were removed.[27]

RO units alone cannot remove lighter molecules like THMs,[28] and nitrates are **not** effectively removed by RO. When RO and GAC are joined in tandem, however, THMs, halogenated organics, many pesticides, and a number of other pollutants can be effectively removed. There are

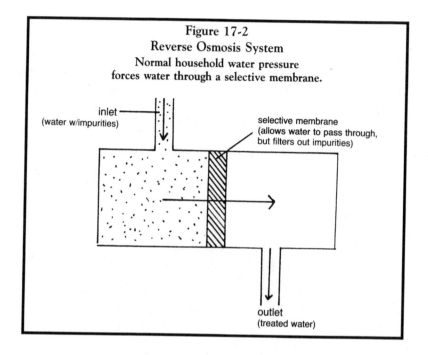

Figure 17-2
Reverse Osmosis System
Normal household water pressure
forces water through a selective membrane.

products manufactured that contain both RO and GAC in one unit. Two RO units with GAC cartridges tested under EPA contract removed more than 80% of non-purgeable halogenated organics.[29] Properly operating RO membranes will remove microbiological contaminants, but leaks or small holes can cause bacteria to get around the membrane and grow to high amounts.

RO units vary widely in price from counter-top models at under $100 to the most expensive undersink models at over $800. The high price is likely the result of the number of accompanying components that are part of an RO system. A prefilter to remove suspended solids will protect the RO membrane and help sustain its life. A GAC filter in the system will make contaminant removal much more comprehensive. Of course the larger the model, the more space must be set aside to accommodate the unit. The consumer should know their own sink area before shopping for a unit. A plumber may be required to install the unit, particularly if it is in-line within the home plumbing system.

Most RO units will need the membrane changed regularly. Again, how often will generally depend on the quality of the water, but could average about once a year. The replacement membrane may cost over $200 depending on the manufacturer.[30] The more expensive units will have built-in backwashing that, according to manufacturers, precludes the requirement of replacing the membrane cartridge. Hot water at a temperature of 145 degrees Farenheit is needed for backwashing. The membrane should be cellulose acetate (CA) if the water is chlorinated, because chlorine will degrade thin-film composite (TFC) membranes.[31] TFC membranes are relatively new, but have proven generally superior to cellulose acetate in removal of organics, and slightly better in removal of inorganics.[32]

RO units do have a few drawbacks. First, they waste a large amount of water. For every treated gallon, four to nine gallons go down the drain. One study of RO units found that as much as 38 gallons are wasted daily.[33] This may be a consideration if the quantity of water available is a concern, as it is in some parts of the country. Also, it may not be economically viable when paying for municipal water. However, this water could be recovered for other uses.

Another drawback is that the system treats the water very slowly. It may take 33 hours to produce five gallons of treated water.[34] As the water is treated it is stored in holding tank to await usage. The treated water may be removed from the holding tank and placed in the refrigerator to allow for continued replenishing of the holding tank.

Look for these features in an RO unit:
- an automatic backflush to clean the membrane and extend its life;
- a prefilter to remove suspended particles that can clog the membrane;
- a warranty on the membrane;
- a method that allows for disinfection of the unit, killing any bacteria that may deposit in the unit;
- specific instructions on when and how to disinfect the unit; and
- an automatic shutoff when the storage tank is full.

Distillation

The distillation process involves separating contaminants by boiling the water to steam and condensing or cooling the purified water into a separate reservoir. Water boils to steam before many heavy organics, inorganics, and metals which are left behind in the boiling reservoir (Figure 17-3). Boiling kills bacteria, viruses, and other microorganisms without question. But, boiling may not be extensive in distillers and none have been advertised as disinfectants.

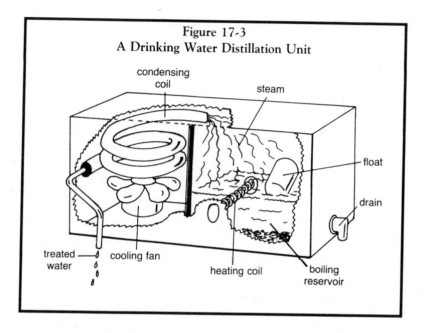

Figure 17-3
A Drinking Water Distillation Unit

condensing coil

steam

float

drain

treated water

cooling fan

heating coil

boiling reservoir

Distillation does not effectively remove volatiles and other chemicals with boiling points lower than or near that of water. These chemicals can follow the water into the finished water reservoir. However, most air-cooled home distillation units have vents to allow some dissipation of volatiles. Water cooled units generally do not allow adequate venting of volatiles. But, some distillation units have a GAC filter in the system which is capable of volatiles removal.

Distillation also softens the water, removing minerals like calcium and magnesium. (Refer to Chapter 6 on the relationship between cardiovascular disease and softened water consumption.) Mineral loss can also make water very aggressive, resulting in the stripping of metals from the plumbing. Consequently, it is best to utilize distillation as a POU device.

This loss of mineral content also changes the taste. Distilled water is generally flat, tasteless, and even bitter.[35] Some people do not care for its taste, so it may be a good idea to try some before purchasing a distillation unit. However, others have found the appreciation of the taste of distilled water to be acquired, and favor it after becoming accustomed to it.

Distillation units come in all sizes and shapes with a variety of features and options. Unit wattages range from 650 to over 1500. The higher the wattage rating, the more water can be treated per hour. You can purchase a unit that must be manually filled and set up, or one that is linked to the plumbing system and is automatic in operation. Automatic models will assure continuous operation of the unit. Be sure to empty the boiling vessel and add fresh water often. Many chemicals, especially inorganics, are concentrated during boiling, increasing the possibility of their movement into the treated water.

One problem with distillation units is the minerals and impurities removed from the finished water can interfere with the operation of the system, requiring frequent cleaning. The frequency depends on the amount of minerals and impurities in the water. Running white vinegar through the distillation unit, much the same way as with cleaning coffee makers, is helpful. Manufacturer products can also be used.

The price for distillation units range from $200 for smaller units to about $2,000 for larger units with a greater capacity.[36] Electric usage will average about 24¢ per gallon depending on your area.[37] If the coolant used to condense the filtered water is itself circulated water, then the water wasted may be more than with reverse osmosis.

Disinfection Treatment Units

If sources of bacteria or microbiological contamination in the well cannot be identified and removed, POU disinfection units can be purchased. These systems include:

Ultraviolet (UV) systems — UV light kills almost 100% of all microbiological organisms in the water passing through the system (Figure 17-4). The unit should come with some type of pre-filter to reduce any suspended material or turbidity. Particulates offer a place for bacteria and microorganisms to hide. UV water treatment units should be equipped with a light sensor, which determines whether the minimum dosage is registering throughout the disinfection chamber, and a shut-off switch or alarm which is activated when the water does not receive an adequate level of UV radiation. Giardia lamblia (see Chapter 8) are fairly resistant to UV radiation, so a five micrometer filter should be installed with the unit, particularly if the source is surface water.[38]

Test the water after installation of the system to be sure it is being disinfected. One study testing the efficiency of three UV units, determined that only one UV was correctly calibrated and considered capable of disinfecting the water under all operating conditions.[39] POU devices should not be used on heavily polluted raw water. An upper limit of 1,000 coliforms per 100 ml of water was suggested for UV devices by the Canadian government.[40]

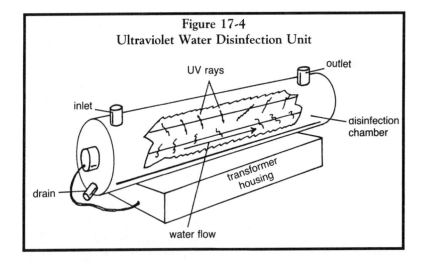

Figure 17-4
Ultraviolet Water Disinfection Unit

Ozonation — An adequate ozone concentration and a sufficient contact time with the water can kill almost all bacteria and inactivate giardia cysts. Viruses are more resistant to ozone than bacteria. A high relative humidity can negatively affect ozone production, reducing the unit's efficiency.[41]

Chlorination — Chlorine can be added to the home's water similar to the way it is added to some municipal systems. POU/POE chlorinators would most likely be used for private wells where THMs and other heavier halogenated organics are not as readily formed due to the low level of natural organics. However, you should test for THMs after installation to be sure they are low. The THM level will also give an indication of the potential for the formation of additional heavier halogenated organics.

Water Softeners

Homeowners, particulary those in hard water areas, are inundated with salespeople attempting to sell them water softeners. Most water treatment companies or dealerships will do minor tests on water, like pH and hardness. (Hardness and pH are explained in detail in Chapters 1 and 6.) While this is often free and may give helpful information, it is usually followed by a very convincing sales pitch to buy or lease their treatment systems.

When dealing with water treatment device salespeople, remember there are no magic cure-alls in this world, and water treatment systems are certainly no exception. Even so, some gear their marketing strategy toward a society that believes there are, and a consumer's lack of chemical knowledge can be exploited. Each water problem is unique, and the solution will not be solved by a one-size-fits-all water softener. While softeners can serve a definite purpose in certain circumstances, claims made by some salespeople can be misleading.

Salespeople usually find prospective customers around the time they purchase a home. If given the go-ahead for a home demonstration, they will likely hook up a portable, faucet-mounted softener to the kitchen tap and go through a series of tests for hardness and iron, using a test kit. These tests will reveal if the water is hard or not, though probably they will find that it is. Most softener companies do not operate in soft water areas. Remember, though, that there are different degrees of hardness.

The author's personal experience with a salesman, officially titled, "water quality specialist," uncovered a number of disturbing claims and inconsistencies:

- He claimed that the unit was not even a softener, but a "special patented treatment device." The dead giveaway was the fact that the unit required periodic replacement of the salt. Any device that removes water hardness and requires the addition of salt is a water softener.
- He claimed that insecticides and virtually all contaminants were removed by this softener, a grossly incorrect claim. More sophisticated treatment devices, like GAC or RO units, are necessary to remove a large range of organic chemicals.
- He claimed that water hardness caused kidney stones, gall stones, hardening of the arteries, and a variety of other ailments. On the contrary, a large number of studies are pointing to soft water, with its lack of mineral content, as contributing to cardiovascular disease (see Chapter 6). Calcium and magnesium, the main elements associated with hardness, are essential elements. Nearly one-fourth of the total calcium intake comes from drinking water, making it an important source of calcium.[42] In fact, many product manufacturers are touting the high calcium content of their products as a benefit to combat osteoporosis. In addition, the EPA is encouraging municipal water systems to reduce acidity and corrosivity by adding calcium carbonate, calcium oxide, or other chemicals to their water to adjust the pH.[43]

- He lacked the technical background to explain the basic operation of the softener and was not clear on the fact that sodium replaces calcium and magnesium in the water. This could be a problem to those reducing their sodium intake on the advice of their physician.
- He claimed that softening the water could save $50 to $60 dollars a month on cleaning costs and reduce wear on appliances. This may be true. Calcium and magnesium, associated with hardness, will "tie-up" detergents and reduce their cleaning power. This generally causes you to use more of the product for the same amount of cleaning. Hardness also produces scale that builds up in hot water tanks, washing machines, and other appliances, thereby reducing their efficiency and lifetime. This is dependent, of course, on how hard the water is. Moderately hard water will not affect the cleaning power as much as very hard water. Therefore the amount of money saved will vary. The unit he was trying to sell costs almost $1200 plus the routine addition of the salt to regenerate the unit's effectiveness.

This example is not an isolated case. **Consumer Reports** recently revealed that similar hard-sell patterns exist nationwide.[44] So, be wary of people using this pitch to sell "special patented water treatment devices."

This is not meant to characterize all water softener companies or their representatives. Many reputable water softener companies exist in the marketplace. But, only your knowledge of the facts will enable you to choose the right product or the right company. It is easy for disreputable salespeople to prey upon a consumer's lack of knowledge. Water softeners can be useful for treatment of hot water or nondrinking water. The decision to invest in a water softener that treats only the hot water would be an economic one.

Choosing the Right Home Treatment Device

Only after testing the water to determine what, if any, treatment is necessary, and after careful comparison shopping, should you purchase a POU or POE device. Get as much information as possible from the salesperson as well as the manufacturer. Do not fall for gimmicks. It is best to deal with reputable, established companies. Check with the Better Business Bureau on any unfamiliar company. Be sure the unit meets the standards of the National Sanitation Foundation (3475 Plymouth Road, P.O. Box 1468, Ann Arbor, MI 48106), an independent laboratory, and the manufacturer belongs to the Water Quality Association (4151 Naperville Road, Lisle, IL 60532, 312-369-1600). The addresses for some POU/POE manufacturers and distributors are listed in Appendix VI.

Emergency Home Treatment Methods

In an emergency situation, there are a few simple treatment methods that can be practiced. These techniques will remove volatile organic chemicals (VOCs) which, as previously mentioned, are most often detected as groundwater contaminants. Be advised, however, that they do not remove the heavier synthetic organics or inorganics and metals. These methods and techniques should be used when alternative water sources are unavailable and it is not possible to remedy the contamination situation, or temporarily until the contamination can be corrected. Of course, the extent of the contamination should be known.

Electric Mixing — Using an egg-beater mixer for at least ten minutes in a stainless steel mixing bowl (do not use plastic) will remove common VOCs very effectively.[45] Use the highest speed setting on the mixer.

Boiling — Rapid boiling is also an excellent method of removing VOCs. Water should be heated to a rolling boil an ' allowed to continue boiling for at least ten minutes. A longer period of time may be necessary to effectively remove some compounds. This is an effective method for killing microorganisms as well. It is, of course, a cumbersome treatment method. You must wait for the water to cool after boiling, and scrubbing to remove mineral deposits from the pan may be necessary. Also, it is considerably more expensive than electric mixing.[46] Time is another a factor to consider when electing to boil water. In addition, nitrates and other organics, if present, will be more concentrated after boiling. When boiling, remember to:

- heat at an aggressive boil *at least* ten minutes;
- use a shallow pan which allows more contact between the water surface and the air; and
- use a hood to remove volatiles in air.

Open standing — Allowing water to sit uncovered in a container for a few days will remove a high percentage of volatiles, but it is not as effective as electric mixing or boiling.[47] One obvious shortcoming is the long wait for water. Another is the possibility of bacterial growth in the water over the standing period. The potential

Table 17-1
Treatment Methods and Alternative Water Sources — Advantages and Costs

Treatment Method/ Alternative Source	Contaminants Removed/Advantages	Estimated Cost
GAC	Removes halogenated and some other organics, radon, & residual chlorine.	Counter top - under $50 and up Undersink - $300 & up plus maintenance costs POE (whole house) - under $1200
Distillation	Removes heavy organics, inorganics, metals, and microbiological contaminants.	$200–$2,000 plus maintenance costs
Reverse Osmosis	Best method of inorganics removal, also reduces the level of many organics, and metals.	$90–$700 and maintenance costs
Bottled Water	Natural spring water from an isolated source could be water of good quality.	$10–$15 per week depending on the size of the household.
Drilling a new well	Possible location of an uncontaminated nearby source.	Averages $10–$15 per foot plus casing and pump costs, nationwide but can range anywhere from $5–$50 per foot depending on the area
Hook up to public water system	Use of water that has been treated by methods specifically designed to remove microbiological contamination.	Usually at least $10,000 depending on the distance from the water main, plus monthly payments.

for this would probably depend on the particular water source.

It is important to be aware of the fact that VOCs are released into the air during the boiling or electric mixing processes, as well as during cooking, showers, baths, and humidifier or vaporizer use. The definite possibility of inhalation of these volatile organics exists. Inhalation of VOCs is certainly an avenue of exposure to reckon with in the same way as ingestion. When utilizing methods like mixing or boiling, it is a good idea to do so outdoors in a well ventilated area, or at least use the hood-fan above the stove.

Alternate Sources of Drinking Water

The framework for assessing water contamination, in the case of an individual well or water system, involves several steps. The first step is to sample and test the water to determine the potential risks involved. If positive results are given, retesting must be done to verify contamination and be certain no additional pollutants are present. Then, based on your perception and interpretation of the risks involved, the decision on whether to treat the water or look into alternate sources must be made. Table 17-1 outlines the costs of treatment methods and alternative water sources.

18. BOTTLED WATER

"When the well's dry, we know the worth of the water."
— Benjamin Franklin

Water is probably the only drink which readily comes to mind that is free of calories, fat, sweeteners, caffeine, and alcohol. Many people want a healthy drink to go with their healthy eating habits and lifestyle, and see bottled water as the pinacle of purity. While the pioneer, Perrier, used to be considered chic, today the main reason bottled water sales are on the rise is the increasing concern about the quality of the water piped into the home.

With Perrier in the lead, bottled water sales have taken off in the last 10 years. In the last five years alone, industry-wide sales have increased 15–20% annually. With the increased demand, bottled water moved from the small specialty shops to the large beverage sections of supermarkets. Industry-wide annual sales are about $2.2 billion.[1]

Over 700 different brands of water are on the market in the United States.[2] Of these, about 80% are not completely natural in their composition and mineral content. These are waters that have been processed, filtered, or distilled. Water that is processed has merely been treated, which consumers can do themselves probably at a much lower cost. According to the International Bottled Water Association, between 25 and 35 percent of the bottled water companies in the U.S. use public water supplies as their source of water.[3] Furthermore, some treatment methods remove necessary minerals. Often a company will add back some minerals after treatment, but they may not be the same amount, type, or in the form they would naturally take for easy digestion or use by the body.

Water can be classified as either **bulk** or **specialty. Bulk water** is usually sold in large quantities and is treated in some way to remove possible contaminants. Water found in water coolers and ordinary tap water are examples of bulk water. **Specialty waters,** such as Perrier, are sold in much smaller quantities as soft drink alternatives. Other classifications of water are:[4]

- *Natural water* — water from a well, spring, or other source where the mineral content is not changed during packaging.
- *Spring water* — groundwater that flows unassisted to a surface opening. Legally, any water carrying the label of "spring water" must be from a spring. A company can use the word "spring" in the company name and not the water name brand itself to get around the requirements. Some spring water may be processed in some way prior to bottling, but *natural spring water* generally means that it was not.
- *Mineral water* — contains dissolved minerals. *Natural mineral water* generally has only minerals natural to the water — minerals it picked up from contact with geologic formations while underground for decades or centuries. Otherwise, a label stating only *mineral* water means the water may have been treated, removing minerals, only to have some added back after the treatment process. Mineral water is usually higher in dissolved solids. The International Bottled Water Association (IBWA) and many states require that to be labeled mineral water it must contain at least 500 ppm of Total Dissolved Solids (TDS).
- *Sparkling water* — is water that is carbonated, which means it contains dissolved carbon dioxide. It is either naturally sparkling, or it is carbonated during bottling. Non-sparkling waters are called *still waters.*
- *Naturally Carbonated Mineral Water* or *Naturally Sparkling Mineral Water* — means the carbon dioxide is from the same source as the water. Generally, waters with the highest content of minerals are carbonated and are common to geothermal regions. In these regions, water from deep within the earth exists at very hot temperatures allowing larger amounts of minerals to dissolve in the water.

- **Processed water** — is artificially carbonated and is usually filtered tap water. Club sodas and seltzers are examples. Club sodas have minerals added while seltzers are low in minerals.

All domestic bottled water in the U.S. is regulated by the FDA which adopts all EPA drinking water standards under the Safe Drinking Water Act (SDWA) of 1974 and its amendments, and applies them to bottled water. Bottled water cannot, however, be granted variances and exemptions from some regulations as public water utilities can in some instances. Some states like California, Florida, and New York regulate bottled water to an even greater extent, requiring testing, mandating additional labeling constraints, and restricting name usage and advertising claims. Pennsylvania requires a report on the geologic conditions and potential pollution near water sources in addition to regular testing.[5]

Since bulk water is generally derived from tap water, it is regulated the same way. Specialty water is assumed to be drunk more selectively than the two liters per day average consumption on which most chemical regulation is based. Therefore it may not be regulated as strictly. Club sodas and seltzers are not classified as bottled water, and are regulated under the category of soft drinks.

Choosing a Specialty Water

Picking a bottled water should not be done casually. The same scrutiny should be applied to any water source, whether from a tap or a bottle. Look for fancy or misleading brand names and avoid them. These may be processed waters, and money can be better spent. Anyone going to the trouble and expense of buying a bottled water wants one that is naturally pure rather than questionably treated.

A lab report on the analysis of the water can be obtained from the company. It is doubtful, to say the least, that a company would release a poor lab report. That would not be good business. However, note whether the tests were performed by the company's own laboratory or an independent lab. If the tests were done by an independent lab, this, of course, lends more credibility to the results. Also, ask about the frequency of testing by the company.

States with stricter testing requirements placed on bottled water companies offer an excellent way to compare lab reports from different companies. Those from more strictly regulated states will have a more comprehensive lab analysis and report. Beware, however, that theoretically (and legally) a bottled water manufacturer could "dump" water that does not pass testing in his state by selling it in another, less regulated state.

The International Bottled Water Association (113 N. Henry Street, Alexandria, VA 22314) may also be a good source of information on a company. The IBWA encourages its members to have their water routinely tested by an unaffiliated laboratory. Membership in the IBWA also increases the company's credibility.

One could even go as far as having the water tested by a commerical lab. Like a well, the quality of water at an isolated spring or other underground natural water source will generally not change appreciably over a period of a few years. A few inexpensive tests will give clues to the source of the water and the validity of the company's claims on the label and in their lab report:

- **Total Dissolved Solids (TDS),** as the name implies, is a measure of all solids dissolved in the water sample. This is, however, also an indirect measure of the mineral level. The higher the TDS, the more minerals present, provided the water is not contaminated.
- **Hardness** is measured as calcium carbonate ($CaCo_3$). Between 75 and 175 ppm is considered hard water. Any more and it is very hard; any less and it is classified as soft. While hardness is principally a measure of calcium and magnesium, like TDS, it is also indicative of mineral content. Usually, the harder the water, the more minerals present.
- **TOX** may be a good test to find chlorinated contaminants. Volatiles, however, will not be accurately represented because most bottles are not completely full and volatiles can escape into the air when the bottle is opened.
- **TOC** gives a good indication of biological contamination. The result should be below 5 ppm.

The lab reports should, of course, not list any SOCs in the bottled water. Tests should report any result as "less than" ($<$) the detection limit for that particular chemical. For example Endrin, a pesticide, should be expressed as < 0.0002 mg/L. Heavy metals and inorganics should also be less than FDA and SDWA standards.

Independent surveys, tests or evaluations are also a good way to pick quality water. Consumer or health magazines regularly evaluate bottled water.

Although nearly all bottled water is treated with ozone as a disinfectant, the bacteriological quality of freshly bottled water varies among different brands, and cannot be guaranteed over the shelf life of the water.

Bacterial counts can increase with storage time. Bottlers often advocate using bottled water in the preparation of baby food and prescriptions, as well as beverages. Therefore, those susceptible to disease, the young, the old and the sick may be opening themselves up to opportunistic microorganisms that could exist in the water. But, putting this in perspective, any beverage or foodstuff may have the same natural inclination for bacterial growth. Refrigeration of bottled water would limit this type of growth.

Taste is obviously an important consideration in choosing a bottled water or in a decision to purchase a home treatment system. Minerals and natural organics determine the taste of water. Pure or laboratory-grade water has no taste, and is not appealing. Good taste, of course, does not mean that the water is chemically safe. New York City and Los Angeles tap waters were determined to be the best tasting waters when compared to 37 bottled waters in a **Consumer Reports** taste test.[6] This is not to say that these or other municipal waters are necessarily inferior in quality to bottled water. Comprehensive testing is required to determine safety regardless of where the water comes from. Generally though, without testing, one who is informed would most likely choose water from an isolated spring rather than from the industrialized Hudson River.

"Plastic taste," as is often experienced when drinking beverages from plastic containers, can be a consideration when purchasing water. Phthalates (see Chapter 12) or other compounds can dissolve into the water from the container. Studies are limited to chemicals leaching from plastic plumbing materials into water, but it may merit concern in beverage containers too as little is known about the effects of phthalates or other chemicals from this source.

When looking at bottled waters, keep in mind that potential sources of water contamination can exist anywhere. Simply because a water is bottled does not mean it is free of contaminants. A study by the New York State Health Department discovered at least one SOC in 68% of 22 bottled waters tested.[7] In 1987, **Consumer Reports** tested selected bottled waters and found that four of them exceeded arsenic and fluoride standards.[8] They also found THMs in six different brands made from tap waters.

One thing to think about is that despite its water contamination problems, the United States generally has rather strict environmental laws and is believed to have better overall water quality than many other nations. At the same time, European standards are much stricter on pesticides and other organics. The THM standard in the U.S. is 100 ppb, while in Europe it is 1 ppb.[9] Therefore, the country of origin of the water could be a consideration in purchasing a bottled water.

While some bottled waters can be a good alternative to contamination problems, it is not the ultimate solution to the quest for safe drinking water. Nor is it a salvation to the continued degradation of water supplies. Although reputable water bottlers go to great lengths to assure purity, manmade contamination can affect bottled water sources just as it can affect any other water source.

19. THE LEGAL GAME

"In law nothing is certain but the expense."

— Samuel Butler (1612–1680)

Anyone who exhibits adverse health effects from toxic chemicals due to the negligence of others has the option of filing a lawsuit against those responsible. This type of lawsuit comes under the classification of **toxic torts.** No provisions for compensation for personal injury or damage as a result of hazardous materials exists under federal law. Those seeking compensation under wrongful trespass, tort law, must look to state laws.

The burden of proof that the defendent's action was a significant factor in causing injury rests on the alleged victim. Under toxic torts, where the plaintiff is seeking compensation for health damage, it must be proven that the chemical exposure is the cause of the health effects. This is not an easy task, considering that many scientists are having trouble determining definite cause and effect relationships for many chemicals in commercial use. Those who experience health effects must prove that they were not caused by other factors or sources. In some cases, however, this has not been necessary. In the case of eight families of leukemia victims that sued three companies in Woburn, Massachusetts for polluting the town's drinking water, the result was an out of court settlement totaling about nine million dollars. While measurably lower than the $400 million asked for, it is one of the largest per-plaintiff settlements for an environmental damage claim.[1] The case never reached the trial stage where medical testimony on birth defects and cancer was to begin, so it was not necessary to prove a cause and effect relationship.

In some instances, reluctance by a company to have the case go to a jury will result in an out of court settlement. By the same token, the prospect of a lengthy and emotionally draining trial for the plaintiff could be the ammunition needed by the defendent to force an early settlement, often for much less than the original figure. Many large firms with equally large assets may not be afraid of the courtroom.

Multiple defendant suits can also be a problem. When more than one party is responsible for hazardous chemical exposure, it may not be possible to prove that any single party contributed to the injury to any measurable degree. In most states, however, if multiple defendants are involved, the plaintiff must prove each was a factor in the causation of illness.[2]

Statutes of limitations had forced plaintiffs to file suit within a specific period of time, usually between one and three years *after the exposure*, depending on the state. The problem being that for many diseases, like cancer, the latency period can be decades. To rectify this, Congress in 1986 set a federal statute of limitations by mandating that any state statute of limitations begin on the date the injury is *discovered.*[3] This is the only federal mandate regarding toxic torts. Legal options and guidelines vary from state to state, including the lengths of statutes of limitations.

Some states, like California, Florida, Minnesota, and New Jersey have moved to enact legislation to make the awarding of compensation to victims of hazardous chemicals a less complicated process.[4] However, while these states have made it easier for injured parties to be compensated, many others have made it more difficult, and have moved to limit compensation. Several states have required:

- plaintiffs to prove liability of each defendant in multiple liability cases;
- limits on punitive and noneconomic damages, thereby limiting the award to victims; and
- limits on attorney's fees which may make it more difficult to secure legal counsel.

According to the GAO, few published decisions on hazardous waste cases exist, generally due to the fact that many cases are settled out of court or remain in litigation.[5] Some courts, however, have permitted parties to seek compensation in the absence of physical injury for mental or emotional distress and impaired quality of life. The New Jersey Supreme

Court in 1987 upheld an award of damages to Jackson Township residents who could not use their contaminated water. The judge ruled that their inability to use water from their wells was an inconvenience and annoyance and therefore a compensable infringement on their quality of life.[6]

A trial can be a long, often frustrating experience with no guarantee of judgment in favor of the plaintiff. For those who have plenty of time, a good case, an experienced lawyer, and the stamina for the long haul, it may prove to be worth the trouble. But, there are no guarantees. No matter how you look at it, the legal game is a tough one. The smooth running "TV trials" are a fantasy. Costs can be minimized by finding a lawyer who will take the case on contingency — a percentage of the settlement plus expenses. At the same time, with these percentages being 30% or higher, and expenses for securing medical, epidemiological, toxicological, hydrogeological, and other information perhaps totalling well over $100,000, the suit could cost much more than anticipated. It also may not be easy to find a lawyer willing to devote a significant amount of time and the initial outlay of expenses to face experienced and well financed corporate lawyers for a longshot case.

In addition to lawsuits for compensation (toxic torts), a citizen can sue to enforce environmental law under the Clean Water Act, the Resource Conservation and Recovery Act, the Safe Drinking Water Act, and the Federal Water Pollution Control Act. In these suits, the plaintiff cannot recover personal damages, but civil penalties, payment of the plaintiff's attorney and litigation costs, halting of pollution, and cleanup costs could be imposed on the violator by the Federal court.[7] This avenue of enforcement of environmental laws is utilized by many public interest groups against corporate polluters. Any citizen can file suit under these federal statutes. The general preconditions of a 60-day notice to the EPA, the appropriate state agency, and the violator, must be met. In the event of inaction by the U.S. EPA and the state agency, a formal lawsuit may be initiated after the 60 day period.[8]

20. MINIMIZING WASTE and POLLUTION

"Ruin is the destination toward which all men rush, each pursuing his own best interest in a society that believes in the freedom of the commons."

— Garrett Hardin, 1968

Water contamination is invariably associated with waste — human waste, chemical waste, trash, garbage, refuse, the throwaways of society. Water draining along the contours of the earth, seeking the lowest possible level, carries with it a myriad of chemical waste into groundwater and surface water reservoirs. Man is the only wasteful animal, and has only himself to blame for the current state of affairs.

In the natural world all resources are necessary and fully utilized to sustain the entire ecosystem. When a predator kills its prey it is because the prey is slowed by disease or old age. The prey population is regulated, securing an abundance of food and their survival. The predator survives as do countless other creatures like birds and insects that feed on what is left behind. What remains is utilized by microorganisms in the soil to aid in the growth of plant life.

Humans, entrenched in a throwaway society, have been, and continue to be, wasteful with their resources, treating them as infinite. Mankind ignores the simple fact that wasted resources are nothing more than environmental contaminants. This includes everything from this piece of paper to chemical wastes from industry. When 100% usage out of a resource is not attained, it becomes a pollutant. Even when waste is properly disposed of by legal or social standards it is still waste, and waste is pollution.

The sheer magnitude of the current municipal waste problem in the U.S. was thrust into the media spotlight with the "Garbage Barge" of 1987. This barge, loaded with over 3,000 tons of solid waste — mostly paper, plastic, and wood — from Long Island roamed the Caribbean for months in search of a state or country that would accept it. North Carolina, Louisiana, Alabama, Mississippi, Florida, the Bahamas, Mexico, and Belize all refused the waste.[1] The incident highlighted and increased public awareness of an ominous fact; landfill space is running out. Landfills are being closed nationwide because they have either reached their capacity or are contaminating groundwater. No new landfills are being developed due to tight government regulation and widespread protests by citizens. Seventy percent of the nation's landfills have closed in the last ten years.[2] An estimated half of the cities in the United States will run out of places to put their waste within the next few years.[3] Many large cities like New York, Los Angeles, Boston, and Philadelphia are nearing the limit of their municipal disposal sites. New Jersey has less than ten municipal landfills still open compared to 100 only a few years ago, and according to Mary T. Sheil, chief of New Jersey's Bureau of Recycling, half of these should be closed.[4] In Pennsylvania the closing of 1,100 dumps and landfills in the last decade has left 82 landfills remaining with an average of about four years of space.[5] Of Ohio's roughly 180 licensed landfills, 50 accept only industrial wastes. Of the remaining 130, forty have under a five year lifespan.[6]

The problem of continually diminishing landfill capacity was spurred by an 80% increase in the waste produced since 1960, due at least in part to the use of heavily packaged products and "convenient" single-serving containers.[7] Roughly half of all household waste is packaging material. Even with a reduction in population, the city of Akron, Ohio picks up nearly 50% more trash yearly from its residents than it did in 1973, exclusive of removal by private haulers.[8] John Steger, Akron's Sanitation Superintendent, points to increases in plastic products, packaging, direct mail advertising, and carryout food as the causes.

Every person in the U.S. produces five pounds of waste daily.[9] Considering most recyclables like paper and aluminum are lightweight, this could be a considerable volume of waste. Everyone contributes to the problem of overcrowded and leaking landfills both directly in the disposal of household waste, and indirectly as consumers of products that generate large amounts of chemical waste as a result of their manufacture. Consequently, all must accept a portion of the responsibility for the mountains of waste, the leaking landfills, and water polluted with industrial chemicals.

Twenty years ago, ecologist Garrett Hardin wrote **The Tragedy of the Commons.**[10] A common pasture available for the grazing of animals owned by local herdsmen had been well kept from overgrazing for centuries. Tribal wars, disease, and poaching kept the human and animal populations at a level where the land could adequately sustain both. When the area achieved social stability, logistics generated the tragedy. Human instinct told each herdsman to try to seek the highest profit. Each herdsman, by adding another animal to his herd, kept all the benefit of the sale of the animal to himself. But, the negative effect of that additional animal on the commons in terms of overgrazing was shared by all the herdsman. Therefore, the personal negative aspect for the herdsman of adding the extra animal was only a small fraction compared to his gain. Seeing only what he could gain, and observing the profits of the other herdsman doing the same thing, each herdsman kept increasing his herd of animals until overgrazing could not sustain the animals any longer and the "Tragedy of the Commons" became a reality.

As Hardin points out, the tragedy of the commons works in reverse in the case of pollution. With the pollution problems of today, human beings add to, rather than take from, the commons. Each human being adds pollutants or waste into the environment. Each person sees the cost of purifying his waste as a significant problem for himself. But, by dumping or discharging it, the cost is spread evenly across the population. The cost of dumping the waste on the commons (the environment) is less than the cost of the individual minimizing or purifying his own waste. All see their own refuse and waste as minimal, but when all this "minimal" waste is added together it has a highly detrimental effect on the commons.

Recycling: Maximum Utilization of Resources

During World War II, the unavailability of foreign sources for the raw materials needed in the war effort made recycling a necessity. A widespread effort by citizens led to comprehensive and successful recycling programs. However, in the post-war years, when raw materials were again available and landfill space was cheap, recycling diminished.

Today, because space in landfills is at a premium and costs are spiralling upward, recycling is re-emerging as a cost-effective disposal method for a wide variety of products and chemical wastes. In the past few years,

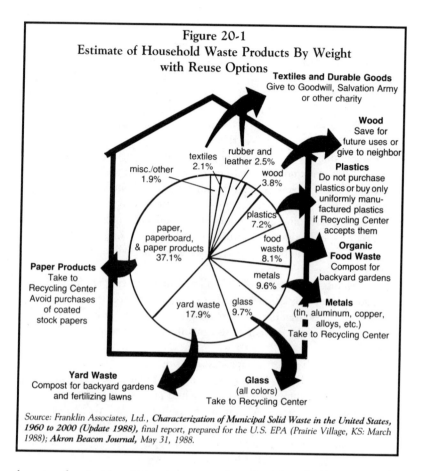

Figure 20-1
Estimate of Household Waste Products By Weight
with Reuse Options

Textiles and Durable Goods
Give to Goodwill, Salvation Army
or other charity

Wood
Save for
future uses or
give to neighbor

Plastics
Do not purchase
plastics or buy only
uniformly manu-
factured plastics
if Recycling Center
accepts them

**Organic
Food Waste**
Compost for
backyard gardens

Metals
(tin, aluminum, copper,
alloys, etc.)
Take to Recycling Center

Paper Products
Take to
Recycling Center
Avoid purchases
of coated
stock papers

Yard Waste
Compost for backyard gardens
and fertilizing lawns

Glass
(all colors)
Take to Recycling Center

textiles 2.1%
rubber and leather 2.5%
wood 3.8%
plastics 7.2%
food waste 8.1%
metals 9.6%
glass 9.7%
yard waste 17.9%
paper, paperboard, & paper products 37.1%
misc./other 1.9%

Source: Franklin Associates, Ltd., *Characterization of Municipal Solid Waste in the United States,
1960 to 2000 (Update 1988)*, final report, prepared for the U.S. EPA (Prairie Village, KS: March
1988); *Akron Beacon Journal*, May 31, 1988.

dumping fees in New Jersey have gone from less than $10/ton to nearly
$100/ton.[11] Not surprisingly, many state and local governments are taking
a serious look at recycling. Several municipalities and the states of New
Jersey, Rhode Island, Connecticut, and Pennsylvania have recently initiated mandatory recycling programs.[12] Still, about 90% of the waste
generated in the United States is buried.[13]

Many still see no reason to break with the tradition they have endured
all their lives of dumping all that they do not desire in one container and
bidding it a less than fond goodbye at the curb. It becomes someone else's
problem. But, it is just this thinking that has brought mankind to where
he is today.

Most people have the "not in my backyard" syndrome, yet most of the garbage they throw out can be reused or recycled, reducing the stress on our natural resources. With the possible exception of certain plastics and other nonbiodegradable packaging materials, nearly all the components of household refuse are recyclable (Figure 20-1). In addition, an amazing amount of energy can be saved and pollution eliminated by the recycling of products routinely thrown out in the trash, such as:

Paper — Recycling saves trees and consequently forests. It requires only 45% to 70% of the energy to make paper from recycled products than from virgin wood pulp.[14] Paper recycling mills cost up to 80% less to build and create significantly less air and water pollution in their processes than a mill designed to make paper from virgin pulp.[15] In addition, five jobs are created when paper is recycled for each job required to manufacture an equal amount of paper from trees.[16] Paper products constitute a broad category: high quality office paper, cards, computer paper, newspaper, cardboard, paper bags, and other assorted papers.

Metals — Recycling lead, aluminum, tin, copper, and other metals ends pollution at both ends of the metals spectrum. It eliminates pollution that occurs from mining the metals' ore and pollution that occurs from disposal of the metal wastes or metal-based byproducts.

It is much easier to reprocess existing metals than it is to start with the metal ore. Reprocessing metals requires less energy and causes less pollution than extracting and processing the metal from its original ore state. Aluminum recycling needs only 5% of the energy and results in only 3% of the water pollution that processing aluminum from its bauxite ore does. Copper recycling requires up to 95% less energy. Recycling steel requires as little as 26% of the energy needed in processing the primary material. In addition, water pollution is reduced by 76%.[17] The U.S. imports 91% of its aluminum yet dumps a million tons yearly, valued at over $400 million.[18] In addition, the U.S. lacks any domestic sources of tin, but fails to get the full use of the metal through recycling.[19]

Glass — Recycling glass and using it in place of nonrecyclable plastics reduces or eliminates the large amount of chemical wastes and pollution created by plastics manufacturing. Glass recycling requires 68% of the energy and produces only 20% of the mining wastes.[20]

Waste Motor Oil — Recycling used motor oil eliminates the pollution caused by its disposal. Oil can be recycled over and over.

To make one quart of motor oil requires almost 17 gallons of crude oil, while 75%–80% of waste oil can be recovered in the re-refining process. Oils refined twice are also generally cleaner. Reprocessed oil also has several other applications, including: cutting oils, lubricants, coolants, transmission fluid, and hydraulic fluid. Approximately 200 million gallons of waste oil is generated just by do-it-yourselfers changing their own engine oil at home. This is a significant pollution source, and only about 15% is recycled rather than dumped in the trash.[21]

Old Tires — There are roughly 200 million tires thrown away annually in the U.S. By simply retreading each once, the demand for synthetic rubber would be reduced by a third.[22] Shredded tires can also be mixed with asphalt for road paving.

Used Lumber — This can be salvaged for non-critical construction uses like tool sheds, mailbox posts, and basement and garage shelving, etc.

Organic Wastes — Food, grass, lawn clippings, and leaves can be composted for backyard gardens.

Paper Recycling

Paper is a widely abused resource. Approximately 45 million tons of waste paper is generated yearly by consumers in the U.S., accounting for about 30% by weight of municipal waste.[23] Excessive use of paper in the packaging of consumer products comprises 65% of the paper used annually in the U.S.[24] Every person in the U.S. is directly or indirectly responsible for about 600 pounds of paper usage per year and 75% ends up in landfills, comprising nearly 40% of the total waste volume.[25]

By recycling every New York Times newspaper printed on any given Sunday, 75,000 trees or roughly 150 acres of forest would remain standing.[26] Yet, each Sunday U.S. citizens throw away 88% of their recyclable newspapers, wasting about a half a million trees.[27] Add this to the countless other paper products which are wasted, the costs of not maximizing this resource begin to reach enormous proportions.

Although aluminum, glass, and steel can be recycled over and over again, paper has a limited life span. It can only make a certain number of passes through the recycling process before fibers become too short and no longer bond together[28] It can however, be mixed with first bond paper.

Full and complete use of every scrap of paper, before recycling it, is something you can do to help reduce the pollutants associated with the pulp and paper industry, the overcrowding in landfills, and the environmental damage associated with forest clearing. There are very few house-holds that really need to go out and buy writing paper. At least one member of an extended family works somewhere where large amounts of scrap paper are thrown away daily. Many letters, memos, publications, and other printed or photocopied material utilize only one side of the paper. This leaves the reverse side for notes, rough drafts, or other non-of-ficial uses. The author wrote all drafts of this book on the backs of old letters and miscellaneous scrap paper.

Incineration of recyclable paper, which many advocate in waste-to-energy plants that burn waste to generate electricity or heat, is not nearly as cost effective as recycling. Waste paper as incinerator fuel is worth only $20 a ton, but recyclable paper is worth $160 or more per ton.[29] Waste energy facilities, which often burn trash without separating or recovering recyclable materials, are a concern because of air emissions associated with these plants.

Plastics Recycling

A significant and still growing portion of household waste, which cannot be easily recycled, is plastics. Plastic bottle usage in the soft drink industry alone rose from relative obscurity in 1975 to hold 69% of the volume of soft drinks marketed in 1985.[30] Other packages and bottlers are following this lead and also marketing products in "convenient" plastic containers. Americans empty plastic bottles at the rate of 2.5 million every hour and recycle practically none of them.[31] According to the plastics industry, not even 1% of the 50 billion pounds of plastic made in the U.S. every year are recycled.[32] Plastics makeup 30% of the volume of waste dumped in landfills.[33] Over 35 states and even some municipalities are looking at legislation to restrict, tax, or even ban some plastic products and packaging.[34]

The problems with plastics are: their manufacture with non-renewable petroleum; their manufacturing processes generate hazardous wastes; they do not biodegrade when discarded; and they generate toxic gases when incinerated. Even if plastics did biodegrade in landfills, the SOC break-down products can contaminate groundwater.

Plastics recycling projects have traditionally not advanced as rapidly as other resources because plastics are not as readily recyclable. The technology to make new plastic containers out of used ones is only now beginning to emerge. Several different types of plastic are in widespread use in a variety of products making the uniform recycling of all plastics a significant problem. However, some plastics can be recycled at similar cost and energy savings as other recyclables. Manufacturing products from scrap plastics requires only 10%–15% of the energy necessary to fabricate the original resin.[35] Used plastics, mainly from plastic beverage containers, are reused for seat padding, sleeping bag and apparel insulation, bath and shower stalls, refrigerator insulation, audio cassette cases, and straps to bind large cartons. Polystyrene can be used to make flower pots, coat hangers, packing material, and insulation.[36] Other plastics can be recylced to be used in the construction of fences, park benches, storage sheds, and other uses traditionally reserved for wood.[37] The Plastics Group at General Electric forsees a nearly endless chain of plastics recycling from pellets, to bottles, car parts, building materials, and components of concrete roadway bases.[38] However, this usage market is limited considering the tremendous amount of plastic bottles and other plastic products manufactured.

Plastic is expected to displace glass as the second most common rigid container in U.S. markets by 1991.[39] According to a ChemSystems report, plastics production will grow just as rapidly in the next 15 years as it has in the last fifteen. The report, commissioned by the Society of the Plastics Industry predicts production to be 76 billion pounds by the year 2000.[40] Despite the limited recycling efforts currently underway, no comprehensive recycling plan exists to keep pace with the increasing rate of plastic production. Even if the recycling of plastics increases due to increased disposal costs, Dr. T. Randall Curlee, an economist at the Oak Ridge National Laboratory in Tennessee, estimates that only 25% of the plastics manufactured in the next 10 years will be separated from the total waste generated.[41]

A number of recycling centers are accepting certain plastic containers. Polyethylene terephthalate (PET) found in soft drink bottles, and high density Polyethylene (PE), found in milk jugs are the plastics in the highest demand by recyclers. These two types constituted 86% of the plastic bottles manufactured in the U.S. in 1986 and are responsible for 64% of all plastics packaging in 1987.[42] Other types of plastic will likely be recycled in the future as petrochemical prices continue to rise.

Generating Demand for Recycled Products

In a free market system, recycling is useless if no one purchases products made from recycled material. Purchasing recycled paper and goods in non-plastic recycled containers sends a clear industry-wide message, creating demand for recycled material. Governments should take the lead in purchasing competitively priced recycled products, like paper, to set an example for their citizens. The state of Maryland has been purchasing recycled paper since 1977 with a total savings of over $17,000.[43] The federal government, on the other hand, has done the opposite. By subsidizing virgin resource usage, they are stifling the market for recycled products. Government subsidies and other economic incentives to companies making paper directly from wood pulp do not allow companies manfuacturing recycled paper to compete on an equal basis.

Industry could manufacture more uniform materials that are easier to recycle, like glass, aluminum, or recyclable plastic. Government on all levels could mandate or increase incentives for recycling. However, citizen action by purchasing recycled and recyclable products rather than throwaways, and by demand for recycling programs is necessary for any meaningful alleviation of a number of environmental and social problems. A widespread committment by citizens to purchase recycled products will force an industry and legislative response. Consumers send mixed signals to industry. Most want environmental problems solved, but still purchase a wide variety of "convenient" products. Non-profit recycling centers exist everywhere and increased voluntary participation by members of the community and a widespread citizen commitment to seek out and purchase products made with recycled material will result in expansion of these recycling programs.

Hazardous Household Waste

In addition to recyclable waste generated by a household, a great number of home-use products contain toxic chemicals (Table 20-1). Most homes, to the surprise of many, are actually small hazardous waste sites. Most people are surprised to find that the chemical products they use regularly are hazardous. Being born and raised in an era of continual product development and marketing, they have come to accept these products with open arms. It is hard for many to look at them as dangerous. While the volume does not approach that of industry, homeowners are collectively the largest single group of hazardous waste generators. A significant amount of hazardous chemicals are purchased, used, and disposed of by the general public. They can be a serious problem if disposed of carelessly. Most of this waste ends up in municipal landfills that were not designed to receive hazardous chemicals. In fact, 20% of the sites on the National Priorities List of Superfund hazardous waste sites are municipal landfills.[44] With millions of homes contributing to this source it has the potential to become a serious pollution problem when all the waste congregates at the local landfill.

Minimizing the waste generated by homeowners and industry is the only sure method of pollution control. All households have a responsibility

Table 20-1
Chemicals in Common Household Products

antifreeze — ethylene glycol

auto electrical system waterproofer— 1,1,1-trichloroethane, methylene chloride

auto engine cleaners — toluene, xylene, trichloroethane, methylene chloride, mineral spirits, petroleum distillates

degreasers — TCE, mineral spirits, chlorinated solvents

dry cleaning fluids — 1,1,1-trichloroethylene, tetrachloroethylene, petroleum distillates

flea powders — (pesticides) carbaryl, dichlorophene, chlordane, methoxychlor

furniture oils & polishes — petroleum distillates

furniture strippers — methylene chloride, methanol, toluene, acetone

gasoline — hydrocarbons, including benzene

nail polish remover — acetone, ethylacetate

oven cleaners — sodium hydroxide, methylene chloride, petroleum distillates

paint thinners — toluene

paint brush cleaners — toluene, naphthalenes, xylene

paints (oil based) — petroleum distillates, aromatic hydrocarbons

stain removers — naphthalene, perchloroethylene

toilet bowl cleaners — hydrochloric acid, chlorophenols

rug cleaners — trichloroethylene

upholstery cleaners — perchloroehtylene, 1,1,1-trichloroethane

Source: Golden Empire Health Planning Center, **Making the Switch: Alternatives to Using Toxic Chemicals in th Home** (Sacramento: GEHPC, 1988).

to reduce hazardous waste. You cannot criticize industry for not taking steps to minimize waste generation if you are not following the same practice in your own home. Avoid buying hazardous chemicals, particularly SOCs. If you feel you must purchase a chemical, do not buy more of the chemical product than you will use. Any unused chemical put out with the trash is waste which becomes pollution.

Some hazardous chemicals can be recycled. Waste motor oil can be put in a plastic container and taken to a recycling center that accepts it. Gas stations will also accept used oil. Auto and other vehicle batteries can be recycled for the lead and other metals. Most gas stations, battery companies, or metal processing companies accept them. Some may even pay cash for them depending on the market value of certain metals at the time.

While the best idea is to buy few chemicals, or none at all, most people have old cans of chemicals lurking in their basement or garage. Among these could be banned organochlorine pesticides or other hazardous chemicals. Banned pesticides should be returned to the manufacturer or retailer. Paint can be given to a neighbor or donated to a civic organization. Other products should be used up or retained in their original container until safe disposal methods are available. Table 20-2 lists some alternatives to routinely used hazardous and nonhazardous household chemicals.

Minimizing Hazardous Industrial Waste

Over 292 million tons of hazardous waste is generated each year in the U.S., with 93% coming from chemical, petroleum, and metals industries.[45] This amounts to over a ton for every citizen in the country. As every industry ultimately caters to the consumer in the form of goods and services, all citizens must share in the responsibility for the hazardous waste generated in the production of the products they purchase. Overseeing the hazardous chemical operations of industry are a variety of understaffed and diverse federal and state agencies which generally leaves effective and safe management of waste by companies to the honor system. Consequently, the citizenry must participate, through legislation and purchasing habits, in the minimization of hazardous waste.

Some companies have greatly reduced the amount of waste they generate and saved substantial amounts of money as well. Cleo Wrap of Memphis, Tennessee the largest gift wrap paper producer in the world switched from organic solvent-based inks to water-based chemical printing inks at a yearly savings of $35,000 in waste disposal costs.[46] Also, because of the elimination of hazardous chemicals, they did not need to store them. Consequently, they could eliminate the use of underground solvent storage tanks and avoid the new federal regulations. Fire insurance premiums were also lowered because they must no longer contend with the fire hazard of stored flammable chemicals.

A 3M electronics plant in Missouri reduced its liquid hazardous waste by 40,000 pounds per year at a savings of $15,000 annually.[47] Company wide, since 1975, 3M reported a savings of $408 million as part of its "Pollution Prevention Pays" program. As a result, 111,000 tons of air pollutants, 15,000 tons of water pollutants, and 388,000 tons of solid waste were not released into the environment.[48]

Despite these and some other success stories and strong incentives

Table 20-2
Alternatives to Household Chemicals*†

Air fresheners: dish of warm vinegar; or
put cloves and cinnamon in boiling water, then simmer; or
set an open box of baking soda or small amount of vanilla in area to be
deodorized.

All purpose cleaner: cleanser and scouring cleanser, baking soda, or Murphy's
Oil Soap.

Ant repellants: Put any of these in areas where you have seen them:
red pepper, chili pepper, paprika, cinnamon, cream of tartar,
salt, dried mint, or sage; or
spread 2 parts boric acid and 1 part sugar mixture around baseboards or
areas you have seen them; or
1/2 cup baking soda/1 T. powdered sugar mixture.

Brass cleaner: make a paste of equal portions of flour, salt, and vinegar. Apply
paste and let dry. Rinse off and polish.

Brass polish: Worchestershire sauce;
apply ketchup, allow to sit, them wipe dry; or
boil onions in water to make a polish.

Carpet cleaner: corn starch and baking soda; or
1/2 cup mild dishwashing detergent in 1 pint of boiling water, whip into a
stiff lather after it cools into a jelly, rub into a small area of the carpet.

Copper cleaner: table salt soaked in vinegar; or
table salt soaked in lemon juice.

Deodorants: use roll-ons, sticks, or creams, NO AEROSOLS.

Detergent: Borax or Arm and Hammer washing soda.

Disinfectants: washing soda and water, Murphy's oil soap, shaving soap, bleach
and warm water; or
Borax and warm water; or
washing soda and warm water.

Drain cleaners: baking soda, then boiling water; or
baking soda, then vinegar, then after 1 hour, boiling water; or
mechanical snake; or
½ cup of salt, then boiling water; or
plunger.

Fabric whitener: Borax or Miracle White.

Flea Repellants: vacuum carpet thoroughly; or
feed pets **small amounts** of brewer's yeast (25 mg for every 10 lbs.),
vitamin B, or garlic; or
buy herbal flea collar and regularly apply oil to it; or
apply rosemary, sassafras, eucalyptus, fennel, or pennyroyal leaves to pet's
sleeping area; or
after bathing, use herbal rinses.

continued

Table 20-2 (cont'd)

Floor Cleaners: vinyl tile or linoleum — 1 cup white vinegar in 2 gallons water, polish with club soda.
wood floors — 1 teaspoon washing soda in 1 gallon hot water.

Furniture polish: make a lemon wax: mix 1 t. lemon oil with 1 pint mineral oil; olive oil;
beeswax; or
rub crushed nuts on wood for oiling polish.

Glass cleaner: ammonia and soap; or
1/2 cup white vinegar in 1 quart warm water;
2 tablespoons lemon juice in 1 quart water; or
baking soda on wet rag.

Gold and silver cleaners: toothpaste

Hair conditioner: yogurt, egg, or olive oil.

Hair gel: dissolve 1 teaspoon unflavored gelatin in warm water, refrigerate until use.

Hair rinses: lemon juice, beer, or vinegar.

Hair spray: 1 chopped lemon (for dry hair use an orange) in a pot with 2 cups water. Boil until 1/2 of water is gone. Allow to cool, then strain. Refrigerate in a pump spray bottle.

Insecticide for houseplants: put soapy water on leaves and rinse.

Mildew stains: bleach;
shower curtains can be washed in 1/2 cup soap and 1/2 cup baking soda with 1 cup white vinegar in the rinse.

Mosquito repellent: plant basil or tansy near doors.

Mothballs: mix 1/2 lb. mint, 1/2 lb. rosemary, 1/4 lb. thyme, and 2 tablespoons cloves and hang in cheesecloth bags;
hang cedar chips, rosemary, mint, dried tobacco, and whole pepper corns mixed in cedar oil; or
store items in cedar chest; hang dried lemon peels, bay leaves, cloves, cedar chips, dried rosemary, and mint, or whole peppercorns; hang dried lavendar; store in airtight container; or spread newspaper around closet.

Oven cleaners: use aluminum foil or trays to protect racks;
put salt on area before it cools, then clean with baking soda and water; or
ammonia* in a dish overnight will loosen burnt-on food, then clean with baking soda and water.

Roach repellent: put chopped bay leaves and cucumber skins where roaches have been seen; or
set out high grade borax or boric acid where roaches are seen and leave for at least 10 days (keep away from children and pets); or
put out equal parts of: 1) oatmeal flour and plaster of paris,
    2) baking soda and powdered sugar, or
    3) borax and brown sugar.
Leaving **no food** (pet food, crumbs, organic food in garbage) around will starve the roaches and force them to "move out."

continued

Table 20-2 (cont'd)

Rust cleaner: Add 1 part cream of tartar to 1 part hydrogen peroxide to form a paste. Apply it to stain and rinse after 15 minutes.

Shaving cream: shaving soap.

Silver cleaner: soak in mixture of:

1 qt. water	1 t. baking soda
1 t. salt	1 piece of cut up aluminum foil; or

water dampened whiting;
wash in mixture of: soap, baking soda, and water; or
rub with toothpaste and a soft cloth.

Snail and slug trap: put stones or clay pots (upside down) near plants, allowing room for snails to crawl under when the weather is hot, remove regularly in the morning and early evening.

Snail and Slug Repellent: put sand or cinders around the base of each plant.

Spot remover: cornstarch or vinegar; or borax in water.
animal stains — baking soda.
blood stains — cold water; soak in 1/4 cup borax and 2 cups water, & wash; or
    rub with paste of cornstarch or cornmeal in water, dry in the sun,
brush; or apply club soda (before stain dries).
coffee — rub egg yolk mixed in warm water on stain.
fruit juice — club soda.
grease spot — Fuller's earth; or rub with borax or damp cloth; or
    make a paste of cornstarch in water, allow to dry and brush.
ink — lemon juice; or mix cream of tartar and lemon juice into a paste, allow
    to sit on the stain, wash; or blot with white vinegar and warm water; or
    soak in homogenized whole milk for several hours.
leather — egg white, beaten.
tea — stretch cloth over the sink and pour boiling water on the stain.
wine — club soda; or use salt as an absorbent; or
    pour salt and hot water on stain and soak in milk.

Toilet bowl cleaner: use baking soda, soap and scrub with toilet brush; or bleach.*

Tub cleaner: white vinegar; bleach* and water; Borax and water; or Borax & lemon juice mixed into a paste, let paste sit before scrubbing.

Window cleaner: 1 T. vinegar in 1 qt. water, water and ammonia.

Wood and furniture cleaner: Murphy's oil soap.

*Never mix bleach or ammonia with commercial cleaners or with each other.
†Children should be kept away from many of these products, including borax or boric acid.

Sources: Cleveland Council on Hazardous Materials; Metropolitan Area Planning Council, A Guide to the Safe Use and Disposal of Hazardous Products (Boston: MAPC, 1982); G. Tyler Miller, Living in the Environment: An Introduction to Environmental Science, 5th Ed. (Belmont, CA: Wadsworth Inc., 1988); Golden Empire Health Planning Center, Making the Switch: Alternatives to Using Toxic Chemicals in the Home (Sacramento: GEHPC, 1988).

for industry to minimize waste — huge increases in the overall cost of hazardous waste management, the financial liability of generating hazardous waste, and the cost of liability insurance — a national waste minimization program does not exist. The U.S. General Accounting Office (GAO) noted that:

> ". . . despite agreement that reducing the generation of hazardous waste is a highly attractive concept and is preferable to controlling waste already generated, both government and industry waste reduction efforts are fragmented and lack a systematic and comprehensive framework for dealing with waste reduction."[49]

Historically, the political, regulatory, and industrial climate has operated exclusively within the confines of waste management. That is, dealing with the waste regardless of the amount, rather than reducing or minimizing it. Only 4% of the hazardous waste generated by industry in the United States was recycled in 1981.[50] Joel Hirschhorn of the Congressional Office of Technology Assessment (OTA) believes that industry could reduce the production of hazardous wastes and pollutants by up to 50% in a few years if industry-wide waste reduction practices were adopted.[51] Both the EPA and Congress support waste minimization efforts, and technologies exist for substantial reduction. However, the EPA does not believe it can make recommendations to Congress on the need to require waste minimization until December 1990.[52]

Even Malcolm Forbes, the late premier capitalist and Editor-in-Chief of *Forbes* magazine, was baffled by the slow progress of corporations who could save billions in disposal costs by transforming wastes into useful or harmless constituents.[53] As he pointed out, corporations spend large amounts of money for acquistion and expansion, yet ignore frugal and very cost-effective ways to minimize waste. Innovative employees who find ways to minimize waste generated by their company will likely be rewarded.

The responsibility of effectively curtailing hazardous waste production should remain with industry, which has the knowledge and capability but lacks the desire. The absence of incentive could be alleviated by economic penalties. Taxing or otherwise fining companies with the capability, but not the willpower, to reduce their hazardous waste would force them to seek ways to recycle or reuse their hazardous waste or minimize the amount they produce. Concentrating efforts on initial disposal costs rather than minimizing waste, only prolongs final payment when costs may be prohibitively expensive.

Currently, the hazardous waste disposal policy in this country is merely stop-gap action rather than a long-term solution.[54] The EPA estimates the cost per metric ton for disposal of hazardous waste under the Resource Conservation and Recovery Act (RCRA) regulations is $90. However, estimates of the cost of cleaning up incorrectly disposed waste is approximately $2,000 per ton.[55] According to the OTA, the future costs will likely be paid by the general public, whereas costs encountered by initial effective management would be more easily and equitably handled by those generating the waste and by those consuming products made from hazardous waste.[56] Waste minimization solves the problem, but current land disposal methods, while initially appearing to be low in cost, are actually adding to the final higher price that must be incurred in the future.

21. THE FUTURE: Altering the Course

"O thou that dwellest upon many waters, abundant in treasures, thine end is come, and the measure of thy covetousness."

— Jeremiah 51:13

In the face of ever increasing chemical usage and mismanagement, concern about your drinking water source is well founded. At the same time, you and your home do not exist on an island, independent of the world and its problems. Pollution of the water resource although a sizable problem itself, is merely a symptom of a much larger one. Clearly, water is not the only resource in jeopardy as a result of an unwavering chemical onslaught and continued misuse of natural resources. Water is only one abused resource in a world of abused resources. Our economic development and technological endeavors have brought us to the stage where we are seeing global scale damage to natural systems that have existed or evolved throughout geologic time. The ability of the earth to exist as a sustainable life support system for *all* its passengers is in doubt as we forge ahead into the next century. The future habitability of the earth is in question.

In his book, ***Living in the Environment,*** G. Tyler Miller lists four progressive stages of awareness of our ecological problems:[1]

1. **Pollution** — discovery of a symptom, like drinking water contamination.
2. **Overpopulation** — the cause of pollution is found to be a large population with wasteful consumption practices.

3. **Spaceship Earth** — finding out that everything is linked to everything else. The Earth and its resources are finite and can be devastated.
4. *Sustainable Earth or Earthmanship* — the realization that to solve ecological problems we must work with nature on the basis of ecological comprehension and stewardship with an emphasis on ethics, because politics, science, and economics have proven insufficient to deal with the crisis.

Despite a recent increase in environmental awareness by many, most people are still in the first or pollution stage, with concern about their immediate needs and resources, like water. Ecological ignorance is widespread throughout the nation and the globe. Humans generally view the world in a society-distorted and simplified way, thinking problems can be easily solved with some quick-fix technology. One widely publicized solution to the catastrophic consequences of ozone layer depletion involved encouraging people to don hats, wear sunglasses, and use sun screen rather than control damaging chloroflorocarbons. Much of the public misses the grave implications of the greenhouse effect; the dramatic changes in climate patterns which could adversely affect the world's food supply, the biological diversity of the planet, the stability of the earth's fragile ecosystems, and ocean water levels. Many people simply say, "if the earth gets warmer, we won't have to go to Florida for warm weather."

What is needed is a greater ecological awareness of, and respect for, the earth's natural ecosystems. Despite our arrogant assumption of advanced intelligence, we know painfully little about them. In this same view, we also must develop a healthy and fearful respect for synthetic chemicals and their hazards, and learn responsible management of our dwindling natural resource base which is necessary to sustain us.

Population Growth

Population growth is probably the most important factor in the generation of human chemical and biological waste and ultimately pollution. Population density directly affects water supplies. The extent of water degradation is proportional to the human population level surrounding it. Too many septic tanks in one area will detrimentally affect the groundwater. Sewage from overcrowded cities must be treated before discharge into waterways so wildlife in the waterway is not devastated and downstream communities receive water that can be readily treated for drinking. A few people living along a river or lake will not affect the quality of the water, even if they dump their human waste directly into the waterway. Water purifies itself quickly when it is not stressed by too much contamination.

The natural cleansing ability of water can handle a small amount of pollution. It is when the human population increases along a watershed system that treatment is necessary to prevent overriding the self-purifying ability of the waterway. The greater the population, the more sophisticated and expensive the treatment system necessary to treat their waste.

While biological contamination is certainly a concern, and has been for centuries, chemical waste contamination is relatively new and has implications for global scale effects. Industries generate products to meet consumer demand and in the process also produce large amounts of chemical wastes. These wastes pollute the air, water, and soil. Consumer demand for new products has been the catalyst for our current waste problems. The more people, the more products manufactured for and consumed by these people, and consequently the more waste generated.

Essentially two types of overpopulation exist. Too many people in a given area for the available life-sustaining resources like food, water, and fuel is one form of overpopulation. This type of overpopulation exists primarily in underdeveloped countries where the population level exceeds the capacity of the land to sustain it. This results in starvation, poverty, and threats to political stability. Bulging populations put a heavy demand on food supplies, while at the same time degrade the land and reduce its food production capability.

In industrialized countries like the United States, another type of overpopulation exists. In this case, people are utilizing resources at a much more rapid rate. We are producing substantially more pollution and causing more damage to the environment than underdeveloped countries which are experiencing high increases in human populations, but comparatively lower rates of resource usage. The population level itself is not a problem, it is the impact the population has on the environment which is the concern. The tremendous amount of pollution per person in this country is the result of the heavy consumption of resources per person. The average person in the U.S. uses 50 times more steel, 170 times more synthetic rubber, 250 times more gasoline and fuel, and 300 times more plastics than the average person in India.[2] If pollution is not adequately restricted, degradation of land and water results in human health effects like cancer and premature deaths.

Unprecedented human population growth has occurred in this century. Well over five billion people now live on this planet and either require or electively consume its resources. The bulging human population in many areas is extending beyond the capacity of resources to support it. It took mankind a long time — until 1850 — to reach a global population of one billion. However, it took only 80 years to reach two billion (1930), 30 years to reach three billion (1960), 15 years to reach four billion (1975), and only 13 years to reach five billion (1988). It is projected that the global population will grow by another billion before the end of the 1990s. This will put additional stress on the earth's natural resources causing reduction in standards of living, increased global pollution, widespread species extinction, and disruption of political and social stability. For these reasons, some see population growth as the single most important problem facing the world.

Changing Attitudes

> *"The dictatorship of humanity has led inevitably to tyranny over other species and over the Earth itself."*
> — Jim Lenfestey 1989

Our present day attitudes and philosophy of throwing away materials after one use is rooted in the past when energy prices were low and it was perceived that a infinite amount of raw materials existed. Also, a much smaller population relied on them. Today, the U.S. generates more solid waste per capita than any other country. As a nation, we drain a precious resource base by generally ignoring adequate management of these resources and comprehensive reuse of waste. We prefer to ravage forests rather than wisely manage wilderness and recycle paper. We favor dumping polluting oil-based chemicals on the ground, into water, and into the air, rather than preserve finite oil reserves by industrial waste minimization and chemical waste recycling.

These wasteful and arrogant practices are based on attitudes that man has carried with him for centuries. Our perception of the world around us has been at least partially determined by our religious and philosophical beliefs. How we view the world is determined by how we see our place in the world. According to historian Lynn White Jr., man's continual destructive drive to dominate nature is deeply rooted in Christian theology.[3] Christianity defined our current concept of progress. The Christian view is that the whole world exists essentially for man's benefit, to be dominated and exploited. Man's progress in the industrial and chemical age fits this attitude. Our political, social, and economic policies exist solely to accommodate and serve humans. However, as White points out, since these attitudes have been with us long before the technological advances of this century, it may be unwise to seek purely technological solutions to our ecological problems. The solution, then, according to

White, must lie in our religious philosophy.

We continually think in exclusively short-range human terms, human suffering and economic loss. We still deal with damage to the earth in economic and social terms rather than on a strictly ecological basis. When forests burn we hear of timber losses. When a large chemical spill occurs on a river we hear of interrupted drinking water service. When government controls curb an industry's emissions, we hear about threatened plant closings and possible loss of jobs. Our concern is whether damage directly affects humans economically in the short term rather than how it affects the global biosphere and man in the long run.

Humans mistakenly believe that they are somehow unrelated to other life on this planet, and can exist apart from the natural world. Many hold a false sense of security wrapped in the artificial shroud that civilized society drapes around them. Even our ongoing references to "the environment" implies it is a separate entity, distanced from ourselves. Because we exist merely as a species in a finite global ecosystem, survival depends on redefining our place in that ecosystem and preserving the land, air, and water for all species. We are the only species which can destroy all others completely, yet we also have the ability to think beyond ourselves and comprehend the logical and ethical implications of preserving the balance of life on the planet.

The natural world operates by its own rules. It does not understand the whimsical forays of human society and rebels against its intrusion. The ecological stability of the earth determines the rules for everything else. Natural laws ultimately have veto power over political, social, and economic issues, whether we as a society realize it or not. In place of the traditional "man-first" philosophy we must view ourselves as merely a part of nature. We must learn to adapt and conform to natural laws, rather than forcing nature to commit to our wishes, or face the catastrophic consequences of this blatant ignorance. In this perspective, we assume ecological responsibility for all our resources. The sooner mankind adopts this ethical respect for the natural world the better our chances to stave off environmental crisis. Preserving the diversity of life and a sound ecological balance on the earth is an absolute requirement for civilizations to endure.

G. Tyler Miller also traces the cultural evolution of man:[4]

1. *Early Hunter/Gatherer Man* — Prehistoric man. Man existed in nature and was controlled by his environment.

2. *Advanced Hunter/Gatherer Man* — Man still existed in nature, made an impact and affected nature, but did not control it. The American

Indian before the Wounded Knee massacre and the Battle of the Little Big Horn would be a latter day example.

3. **Agricultural Man** — Ten to twelve thousand years ago. Man began to cultivate the land using advanced tools. Man moved from existing in nature to combating it, and through advanced agricultural methods increasingly began to control it.

4. **Industrial Man** — Man remains against nature, but has great control over it. He discovers the secrets of tapping fossil fuels for energy and synthesizing chemicals. But, widespread pollution, increased population growth, and resource exploitation and depletion brings about the realization that there are limits to his control over nature.

5. **Sustainable Earth or Earthmanship Man (man in a sustainable global society)** — Man working with nature in a cooperative ecological effort. An earthmanship or sustainable earth view and attitude by the earth's inhabitants would treat the disease and not just the symptoms (i.e., water pollution and the contamination of drinking water) of our environmental problems.

Our current state of perpetual environmental degradation will not change until we alter the attitudes and thinking that has brought about the current state of affairs. According to Miller, because population levels are so high and technology is so ominous, we may have as little as 50 years to make the transition to a sustainable earth society or an undesirable change will be thrust upon us.[5]

Technology and Progress

"Mistrust of technology is an attitude that ought to be taken seriously. It has positive value in avoiding grave disasters."
— Roberto Vacca 1974

Many people believe that humans can artificially sustain themselves indefinitely using advanced technologies; that we can separate ourselves from the environment and the problems we have created. This goes squarely against natural laws to which we are ultimately and irrevocably bound. Blind and unquestioning confidence in unchecked technological development has put us on the precarious precipice where we exist today. Even with the tremendous technological advances in this century, human poverty has increased while ecological integrity has drastically decreased. Current technologies, although widely available and comprehensive in addressing numerous immediate needs of mankind, lack an ecological foundation and sustainable direction. This is not to say that technology itself is bad, sustainable technologies utilizing renewable resources hold much promise. Technology will not be the downfall of mankind, but continued mismangement and misdirection of technology certainly will.

In view of the sweeping and ominous potential for global damage, a fundamental reevaluation of our concept of progress is necessary. Traditionally, it has been a nondirectional progress simply for the sake of progress without regard to ecological considerations even on the smallest local scale. Nearly all development and technological advances are unconditionally welcomed, almost worshipped in wide-eyed wonder of human capability despite the growing legacy of damage caused by the irresponsible deployment of many of them. Because a specific technology exists does not automatically imply it should be developed. Careful evaluation of its ecological advantages must be accomplished first. Often a new technology is developed for the purpose of repairing the negative effects of an earlier technology.

Just as our scientific knowledge and assessment of the human health effects of localized pollution are inadequate, so is our ability to determine the effects of human activity on the world environment. Monitoring efforts to measure industrial man's damage to his world lack commitment and expertise on the international scale. Our limited knowledge is not capable of foreseeing the far reaching ramifications of the many seemingly innocent actions we take with respect (or lack of respect) to our environment. Today we are paying for the ignorance of the past; tomorrow we will pay for the uncertainty of today. This can be assured in a high-tech society hurling itself forward without an adequate braking system. Our ability to deal ecologically with the negative aspects of technology and development does not parallel our ability to create them. Technological development has left us with a tremendous hazardous waste problem. New dangers to match the likes of PCBs, dioxin, DDT, and chlorofluorocarbons, are most certainly looming on the horizon. Careful planning is necessary to assure no problems manifest themselves in the future as a result of our endeavors today. Living in a technological age requires technological responsibility and accountability.

Economics, Resources, Society, and Solutions

> *"Most of the luxuries, and many of the so-called comforts, of life are not only not indispensable, but positive hindrances to the elevation of mankind."*
>
> — Henry David Thoreau (1817–1862)

Many people fail to understand how closely related and dependent the economy and our natural resources are on each other. Both are often viewed as mutually exclusive. The natural world is, in fact, quite similar to the economy in many respects. Both are extremely fragile, misunderstood, unpredictable in their response to human endeavors, and subsist on the brink of disaster.

It is mistakenly believed that to further economic interests natural resources must be sacrificed, and to protect resources would be to invite economic ruin. Jobs are continually at odds with environmental preservation in political arenas. A current example is the battle over the future of old-growth forests. The logging industry maintains that old-growth forest ecosystems must be sacrificed for the health of the industry and local economies. But, any economy whose existence depends on the destruction of a finite ecosystem cannot be a strong economy in the first place.

Economic growth is perpetual only as long as the resource base can support it. This cannot occur when increasing population levels and irresponsible management destroy the basic resources necessary to fuel that growth. Global economic growth has increased in the past four decades, but finite natural resource systems that feed and sustain it have not. The economy, and consequently our quality of life, can be seriously crippled or even devastated by continued mismanagement of our natural resources. Our environment is the foundation of any and all economic growth and enterprise. The economy is not based on technology but on resources. The effective management of these resources is the basis for long-term economic growth. Proper management of a resource base is containment of these resources within the system; a closed system where resources are wisely managed and continually reused. The compromise of resources in favor of economic growth will ultimately result in the breakdown of both the resource base and the economy.

History has shown us that the survival of any civilization is completely dependent on the availability and effective management of resources. One of the main reasons for the decline of the great Babylonian empire was massive deforestation. The subsequent soil errosion caused silt to choke the Tigris and Euphrates Rivers, seriously crippled agricultural production.[6] It is believed that the ancient Mayan civilization of Latin America overextended its resource base leading to the breakdown of its agricultural system. This, in turn, lead to a shortage of food and the collapse of its social structure.[7]

Is our high-tech industrial society shielded from such a fate? Will we continue to misuse our precious resources or learn to fully utilize them through reuse, conservation, and preservation? We have treated environmental protection as some sort of luxury, rather than a means of survival. Forests worldwide have been completely decimated for the quick profits from timber sales and the short-term value of soil for agriculture. The long-term economic potential of innumerable forest products that can be realized through sustained preservation of the resource has been ignored. According to the Worldwatch Institute, a non-profit international environmental watchdog, countries that can manage their resources effectively in a closed system of reuse will be the ones to thrive economically in the future.[8] Countries that have squandered their natural resources and lack enough to sustain their large populations will not.

The solutions to our environmental problems are quite simple in their ecological theory, but very difficult within the socioeconomic and political constraints we have placed on ourselves:

- We must *make a transition from the use of polluting fossil fuels,* such as oil and coal, *to the utilization of renewable solar and wind energy sources.* The Reagan administration, in its eight years in power, cut the budget for energy conservation by 89%. Funds for research on alternate energy sources were slashed by 83%.[9] It chose instead to chase down fossil fuel reserves worldwide, apparently oblivious to the political and environmental consequences. By utilizing renewable energy sources, the widespread pollution from fossil fuel exploration, transportation, processing, and combustion can be significantly reduced or eliminated. In addition, at 1986 production rates, the world's oil supply has only about 33 years remaining.[10]

- We must *bring the global population under control through international family planning.* The more people, the more waste is generated. This puts more stress on resources resulting in the decline of civilizations.

- We must *halt the current trend of massive deforestation* which has perilous global consequences. This includes protection of existing forest areas and the initiation of large scale replanting programs in the U. S. and throughout the globe. In just 30 years, nearly 50% of the world's tropical rainforests have been cut down for lumber or agriculture. According to the World Bank, deforestation in the Amazon encompassed 372,600 square miles.[11] Almost 20 million acres of tropical forest, an area the size of Massachusetts, Vermont and New Hampshire combined, are lost every year.

The problem is not restricted to the tropics. U.S. government and industry are also guilty of wasteful and damaging timber cutting practices. The U.S. Forest Service managed to lose over two billion dollars in the below-cost sale of timber in the past ten years.[12] In 1986 alone, the federal government lost $600 million in road building projects through wilderness areas and in management of timber sales.[13] The logic behind the U.S. Forest Service's policy is the preservation of the economy surrounding the timber industry. This, however, occurs at the expense of recreation, paper recycling, and natural resource conservation, all of which employ more people than the timber industry.

Deforestation is not merely the loss of a few trees. Aside from the threats to species diversity or extinction for a large number of species, deforestation increases carbon dioxide, which speeds up global warming, and disrupts the hydrologic cycle which determines global weather and climate patterns. The clearing of forests also causes soil runoff, which reduces regrowth capability and chokes wildlife in streams. It also removes the shade over streams, increasing water temperature beyond the tolerance level for certain species of fish and aquatic life.

- *We must minimize the generation of all waste, particularly hazardous chemical wastes.* All that we now consider to be waste must be treated as a resource to be utilized by a global society wishing to sustain itself indefinitely.
- *We must accept that everything is connected to everything else.* This necessitates a humble respect and commitment to harmonious co-existence with the natural world as opposed to the attitude of relentless supremacy. We do not know, nor will we ever know, everything about the global ecosphere and its intricate web of life.

These solutions may require personal sacrifice, a shift in economic priorities, and global unity and cooperation. We have created these ecological problems, so it seems reasonable to assume that we can solve them. But, we must have the willingness to solve them. Not many people realize what this entails and how we will have to alter our lives. It will be difficult to convince the ecologically ignorant of the need for these changes.

Future generations will likely curse our disregard for their world; the way we relentlessly and thoughtlessly continued to damage the earth despite its warning signs. They will bear the brunt of our arrogance and ignorance if our present course is not significantly altered. However, the immediate costs of not dealing with booming population growth and the pace of unchecked technology will be the loss of our personal and economic freedoms. Governments will find it necessary to control pollution and resource depletion through regulation. These restrictions will directly impact our quality of life and determine how we live, work, and travel.

What You Can Do

People generally have a habit of avoiding or denying a problem until it reaches mammoth proportions. Many generally ignore all problems that do not immediately concern them, and will utilize resources based on their economic status. Most people, if asked, would say they are all for protecting the environment, but want someone else to do it. Few fully comprehend what is fundamentally necessary to attain ecological stability and sustainable societies. Many are wrapped up in the seemingly isolated corner of the world that is their lives. They refuse to see the devastating effects that will result from a lack of commitment to solving our environmental problems.

Awareness of our ecological problems is increasing. This is due to the fact that some problems are beginning to manifest themselves: global warming; a shortage of waste disposal sites; increases in groundwater contamination incidents; depletion of the ozone layer; and increasing evidence of the adverse health effects of pollution. Concern for the environment is "in vogue." But it is not trendy or simply another political issue. It is the whole ball of wax, all the marbles. The push and pull of our modern bureaucratic system and the relentless, single-minded drive of industry leaves the protection of our natural environment in the hands of the people. We will all reap the benefits or suffer the consequences.

We are now entering the last decade of a century of unprecedented growth and technological development. As the global environmental crisis worsens, the 1990s will be the decade where we will be forced to reassess population growth, energy policies, development, and chemical and technological progress as they relate to the integrity of natural global systems. It will be a crucial time in the evolution of mankind. We must either change the way we view our place within the global ecosystem or risk a disasterous spiral from which we may never recover. Every generation has faced an uncertain future, however, none has had to deal with the highest stakes of all — the survival of the earth itself. Choices must be made, and changes must result. These changes can only occur through active participation by all passengers on this planet. There are several actions you can take that can make a difference.

- Buy natural products like wood, cotton, and wool, and purchase only containers of glass, aluminum, tin, and cardboard. Avoid purchasing nonrecyclable plastic.
- Recycle everything you possibly can. When you are discarding anything, ask yourself where it might be utilized. Support local recycling efforts and legislation. Do not be concerned with whether you will make money on your recycling. Do it because it makes sense.
- Get the fullest possible use of nonrenewable and minimally recyclable products. Write on both sides of the paper and use that scrap wood. Simply because our society gives the illusion of free-flowing resources, does not mean they should not be utilized to the maximum extent possible. Resources are too often taken for granted and misused.
- Be selective and minimize your purchases of wood products like paper and lumber. Because timber resources are widely available does not mean the trees were cut in an ecological manner to sustain forest ecosystems. Current government and logging policy favors a few jobs over the ecological and recreational value of our forest lands, including virgin and old growth forests.
- Be prudent about the products and packaging material you purchase as not all waste can be recycled. Do not buy anything made from foam or styrofoam. Everyone must be accountable for their waste, so plan accordingly. Think about where it will go and how it will be disposed of when you purchase something. If there is no place to recycle it, do not buy it.
- Purchase products made from recycled material whenever possible. (See Appendix VII.)
- Plant *native American* trees and see that they grow. Protest the cutting of any healthy trees in your community. Trees are often sacrificed in the name of a new curb, underground pipeline, or other triviality. Even dead trees are needed by wildlife.
- Minimize your use of electicity. Utilize natural light, or change to energy-saving lights.
- Upgrade the insulation in your home. Insulate your hot water heater.
- Turn your thermostat down and put on a sweater.
- Drive at 55 miles per hour on the freeway. Elsewhere, drive at a steady smooth speed. Anticipate stops and avoid jack rabbit starts. Drive as if you have a glass of water on the dashboard.

- Keep your car and air pollution equipment in good condition.
- Take public transportation, ride your bicycle, or walk instead of using your car.
- Obtain as much of your energy as possible from renewable sources like sun and wind. Solar water heaters are cost-efficient over lifetime use. Solar electricity (photovoltaic cells) will be cost-effective by the mid-nineties.
- Compost household food wastes for an organic garden in your backyard.
- Join with others in the cause of global ecology by joining and supporting national environmental organizations. (See Appendix VIII.) There is strength and power in numbers.
- Make conservation part of your everyday life. Set an example for others. Teach your children ecological and conservation principals. Every little bit helps when everyone does as much as they can, just as every little bit causes damage when everyone is involved.
- Discourage over-development in your area.
- Encourage your elected government representatives to enact legislation that:
  - increases incentives for industries to minimize waste;
  - encourages a shift to renewable sources of energy like wind and solor power;
  - initiates solid workable programs to stabilize the global human population level;
  - requires banks lending money to nations and businesses to ensure that the project results in not only environmental protection, but environmental restoration;
  - makes the production of damaging chemicals, like CFCs, illegal; and
  - requires increases in energy-efficiency and encourages energy conservation.

It will not be enough to talk about it. Only by action, before it is too late, can the earth be replenished and maintained as a viable support system for all its inhabitants. This accomplished, incidents of drinking water contamination will be a thing of the past, rather than on the increase.

References

2. Living in a Chemical World
[1]Samuel S. Epstein, *The Politics of Cancer* (Garden City: Anchor/Doubleday, 1979), p. 30; U.S. International Trade Commission, *Synthetic Organic Chemicals, U.S. Production and Sales 1988* (Washington, D.C., September 1989).
[2]Epstein, p. 11.
[3]American Cancer Society, *Cancer Facts and Figures - 1987.*
[4]E. Pollack and J. Horm, "Trends in Cancer Incidence and Mortality in the U.S., 1969–76," *JNCI,* 64 (1988), 1091.
[5]Toxic Substances Strategy Committee, *Toxic Chemicals and Public Protection,* Council on Environmental Quality (Washington, D.C.: U.S. Government Printing Office, 1980).
[6]Samuel Epstein and Joel Swartz, "Carcinogenic Risk Estimation," *Science,* 20 May 1988, p. 1043–1045.
[7]Ibid, pp. 1044–1045.
[8]Michael Castleman, "Toxics and Male Fertility," *Sierra,* March–April 1985, p. 49-52.
[9]Manning Feinleib and Ronald W. Wilson, "Trends in Health in the United States," *Environ. Health Persp.,* 45 (1985), 267–276.
[10]National Cancer Institute, "Cancer Control Objectives for the Nation: 1985–2000," *NCI Monogaphs,* No. 2 (1986), p. 15.
[11]Samuel Epstein and Joel Swartz, "Fallacies of Lifestyle Cancer Theories," *Nature* (London), 289, 15 January 1981, 127–130.
[12]William J. Blot, et. al., "Cancer Mortality in U.S. Counties with Petroleum Industries," *Science,* 7 October 1977, 51–53.
[13]Marise S. Gottlieb, Charles L. Shear, and Daniel B. Seal, "Lung Cancer Mortality and Residential Proximity to Industry," *Environ. Health Persp.,* 45 (1982), 157-164.
[14]John Kaldor, et. al., "Statistical Association between Cancer Incidence and Major-Cause Mortality, and Estimated Residential Exposure to Air Emissions from Petroleum and Chemical Plants," *Environ. Health Persp.,* 54, 1984, 319–332.
[15]Natural Resources Defense Council, *A Who's Who of American Toxic Polluters* (Washington, D.C.: NRDC, June 1989).
[16]Robert Hoover, et. al., "Cancer By Country: New Resource For Etiologic Clues, *Science,* 19 September 1975, p. 1005-1007.
[17]*National Center for Health Statistics: Health, United States, 1988,* DHHS Pub. No. (PHS) 89-1232, Public Health Service (Washington, D.C.: U.S. GPO, March 1989).
[18]Robert Hoover and Joseph Fraumeni Jr., "Cancer Mortality in U.S. Counties With Chemical Industries," *Environmental Research,* 9 (1975), 196-207.
[19]Michael Kennedy, "By 'Old Man River,' New Health Fear," *The Los Angeles Times,* 9 May 1989, p. 1 & 20.
[20]Council on Environmental Quality, *Chemical Hazards to Human Reproduction* (Washington, D.C.: U.S. GPO, 1981).
[21]Ibid.
[22]"Cause For Concern and Optimism," *Newsweek,* 16 March 1987, p. 63-66.
[23]J.G. Wilson, *Environment and Birth Defects,* (New York: Academic Press, 1973).
[24]W.D. Mosher, "Reproductive Impairments in the United States 1965–1982," *Demography,* 22, No. 415 (1985), 430.
[25]National Academy of Sciences, *Drinking Water and Health,* Vol. 6, Safe Drinking Water Committee, National Research Council (Washington, D.C.: National Academy Press, 1986), p. 37.
[26] B.J. Dowty, J.L. Laseter, and J. Storer, "The Transplacental Migration and Accumulation in Blood of Volatile Organic Constituents." *Pediatric Research,* 10 (1976), 696–701.

[27]K. Hemminki, et. al., "Spontaneous Abortions Among Female Chemical Workers in Finland," *International Archives of Occupational and Environmental Health,* Vol. 45 (1980), pp. 123–126.

[28]K. Hemminki, et. al., "Transplacental Carcinogens and Mutagens: Childhood Cancer, Malformations, and Abortions as Risk Indicators," *J. of Toxicol. & Env. Health,* 5 (5,6), 1115–1127.

[29]O. Merrik, et. al., "Major Malformation in Infants Born of Women Who Worked in Laboratories While Pregnant," *Lancet ii* (1979), p. 91.

[30]F. Funes-Cravioto, "Chromosome Abberrations and Sister-Chromatid Exchange in Workers in Chemical Laboratories and a Rotoprinting Factory in Children of Women Laboratory Workers," *Lancet ii* (1977), pp. 322–325.

[31]Castleman, p. 49.

[32]Ibid.

[33]John Elkington, *The Poisoned Womb* (New York: Viking Penguin, 1986), p. 120.

[34]Stanley J. Kabala, "Poland, Facing the Hidden Costs of Development," *Environment,* 27, No. 9 (1985), 10.

[35]Ibid, p. 9.

[36]G. Tyler Miller, *Living in the Environment,* 5th ed. (Belmont, CA: Wadsworth Publishing Co., 1988), p. 13.

[37]Jean Pierre Lasota, "Darkness at Noon," *The Sciences,* July/August 1987, pp. 23–29.

[38]Ibid, p. 24.

[39]Committee Poland 2000, *Prognosis of the Health of the Polish Society at the Turn of the Twentieth Century* (Warsaw: Polish Academy of Sciences, 1983).

[40]Kabala, p. 12.

[41]James Bovard, "Headlights at High Noon, A Silent Spring in Eastern Europe," *New York Times,* 26 April 1987, p. F3.

[42]Joann S. Lublin, "Alzheimer's Linked to Aluminum in Water," *Wall Street Journal,* 13 January 1989, p. B2.

[43]Daniel P. Perl, "Relationship of Aluminum to Alzheimers Disease," *Environ. Health Persp.,* 63 (1985), 149-153.

[44]Floyd Taylor, et. al., *Project Summary: Acid Precipitation and Drinking Water Quality In the Eastern United States,* EPA 600/S2-84-054 (Washington, D.C.: U.S. EPA, April 1984).

[45]Roger Lewin, "Parkinson's Disease: An Environmental Cause?" *Science,* 19 July 1985, p. 257-258.

[46]Richard Clapp, et. al., "Leukemia Near Massachusetts Nuclear Power Plant, *The Lancet,* 5 December 1987, pp. 1324–1325.

[47]*National Center for Health Statistics,* p. 21.

[48]Ruth A. Lowengart, et. al., "Childhood Leukemia and Parent's Occupational and Home Exposures," *JNCI,* 79, No. 1 (1987), 39-46.

[49]Ibid.

[50]Lester R. Brown, *State of The World 1988,* Worldwatch Institute (New York: W.W. Norton and Company, 1988), p. 15.

[51]A.E. Strong, "Greater Global Warming Revealed by Satellite-Derived Sea-Surface-Temperature Trends," *Nature,* 338, 20 April 1989, 642–645.

[52]Christopher Flavin, "Slowing Global Warming," *Env. Sci & Technol.,* 24, No. 2 (1990), 170–171.

[53]Ibid.

[54]Richard A. Kerr, "Global Warming Continues in 1989," *Science,* 2 February 1990, p. 521.

[55]Brown, p. 4.

[56]Ibid.

[57]Carl Sagan, "A Piece Of the Sky Is Missing," *Parade,* 11 September 1988, pp. 10-15.

[58]G. Tyler Miller, p. 499.

⁵⁹U.S. Congress Office of Technology Assessment, *Are We Cleaning Up? 10 Superfund Case Studies — Special Report,* OTA-ITE-362 (Washington, D.C.: U.S. GPO, June 1988).

⁶⁰Frank Lautenberg and Dave Durenberger, "Lautenberg-Durenberger Report on Superfund Implementation: Cleaning Up the Nation's Cleanup Program," Senate Subcommittee on Superfund, Ocean and Water Protection, May 1989, p. 185.

⁶¹Ibid.

⁶²"Summary of Subcommittee Groundwater Monitoring Survey," Subcommittee on Oversight and Investigations, Committee on Energy and Commerce, U.S. House of Representatives, April 1985.

⁶³Environmental Defense Fund et. al., *Right Train, Wrong Track: Failed Leadership in the Superfund Cleanup Program,* June 1988.

⁶⁴Ibid, p. 2.

⁶⁵Lautenberg and Durenberger, p. 21.

⁶⁶Ibid.

⁶⁷Ibid, p. 38–39.

⁶⁸Ibid, pp. 38–40.

⁶⁹U.S. Congress Office of Technology Assessment, *Technologies and Management Strategies for Hazardous Waste Control* (Washington, D.C.: Congress of the United States, September 1983).

⁷⁰Statement of Richard L. Hembra, Director Env. Protection Issues Resources, Community and Economic Development Division, before the Subcommittee on Oversight and Investigations of the U.S. House Committee on Energy and Commerce, In: U.S. General Accounting Office, *Groundwater Conditions at Many Hazardous Waste Disposal Facilities Remain Uncertain,* GAO/T-RCED-89-30 (Washington, D.C.: U.S. GAO, April 27, 1989).

⁷¹Ibid.

⁷²U.S. EPA, *Surface Impoundment Assessment National Report,* EPA 570/9-84-002 (Washington, D.C.: Office of Drinking Water, December 1983).

⁷³Ibid.

⁷⁴Ibid.

⁷⁵Ibid.

⁷⁶U.S. General Accounting Office, *Report By the Comptroller General Of the United States: EPA's Inventory of Potential Hazardous Waste Sites Is Incomplete,* GAO/RCED-85-75 (Washington, D.C.: U.S. GAO, March 1985).

⁷⁷U.S. General Accounting Office, *Superfund: Extent Of Nation's Potential Hazardous Waste Problem Still Unknown,* GAO/RCED-88-44 (Washington, D.C.: U.S. GAO, December 1987), p. 13.

⁷⁸U.S. GAO, *Hazardous Waste: Uncertainties Of Existing Date,* GAO/PEMD-87-11BR (Washington, D.C.: U.S. GAO, February 1987).

⁷⁹U.S. OTA, *Are We Cleaning Up?,* p. 13.

⁸⁰Epstein, p. 27.

3. Water Pollution & Drinking Water Contamination

¹Conservation Foundation, *State of the Environment: A View Toward the Nineties* (Washington, D.C.: Conservation Foundation, 1987), p. 23.

²U.S. House Committee on Merchant Marine and Fisheries, *Coastal Waters in Jeopardy: Reversing the Decline and Protecting America's Coastal Resources,* Oversight Report, (Washington, D.C.: U.S. GPO, 1989).

³Ibid, p. 1.

⁴Tim Smart, and Emily Smith, "Troubled Waters: the World's Oceans Can't Take Much More Abuse," *Business Week,* 12 October 1987, p. 98.

⁵U.S. Office of Technology Assessment, *Wastes in Marine Environments,* OTA-0-334 (Washington, DC: U.S. GPO, April 1987), p. 186.

⁶Linda S. Sheldon and Ronald A. Hites, "Organic Compounds in the Delaware River," *Environ. Sci. Technol.,* 12, No. 10 (1978), 1188–1194.

⁷R.J. Kuzma et.al., "Ohio Drinking Water Source and Cancer Rates," *Am. J. Pub. Health,* 67, No. 8 (1977), 725–729.

⁸U.S. EPA, "EPA States Plan Cleanup of U.S. Waters Plagued by Toxics," *Environmental News* (Washingotn, D.C.: EPA Office of Public Affairs, June 1989).

⁹Philip Shabecoff, "U.S. Says Toxic Chemicals are Imperiling Aquatic Life," *The New York Times,* 14 June 1989, p. A24.

¹⁰U.S. OTA, p. 197.

¹¹Charles Kupchella and Margaret Hyland, *Environmental Science: Living Within the System of Nature* (Boston: Allyn and Bacan Inc., 1986), p. 374.

¹²U.S. General Accounting Office, *Water Pollution: Improved Monitoring and Enforcement Needed for Toxic Pollutants Entering Sewers,* GAO/RCED-89-107 (Washington, D.C.: U.S. GAO, 1989).

¹³U.S. OTA, p. 204.

¹⁴U.S. General Accounting Office, *The Nation's Water, Key Unanswered Questions About the Quality of Rivers and Streams,* GAO/PEMD-86-6 (Washington, D.C.: U.S. GAO, September 1986).

¹⁵Greenpeace, *Water for Life, The Tour of the Great Lakes on the Beluga* (Washington, D.C.: Greenpeace, 1989).

¹⁶Rudolf Jaffee and Ronald A. Hites, "Fate of Hazardous Waste Derived Organic Compounds in Lake Ontario," *Environ. Sci. Technol.,* 20, No. 3 (1986), 267–274.

¹⁷Jon R. Luoma, "Damaged Wildlife Shows Pollution Still Plagues Great Lakes," *The New York Times,* 12 July 1988, p. C4.

¹⁸Charles E. Cobb, Jr., "The Great Lakes Troubled Waters," *National Geographic,* July 1987, pp. 2–31.

¹⁹Norman L. Dean, *Danger on Tap, The Governments Failure to Enforce the Federal Safe Drinking Water Act* (Washington, D.C.: National Wildlife Federation, October 1988).

²⁰"Study of Drinking Water Assails EPA as Derelict in Monitoring," *The New York Times,* 6 January 1988, p. A14.

²¹Joe D. Francis, et. al., *Executive Summary, National Statistical Assessment of Rural Water Conditions* (Cornell University, Department of Rural Sociology, 1984).

²²Veronica Pye and Jocelyn Kelley, "The Extent of Groundwater Contamination in the United States," In: *Groundwater Contamination,* National Research Council (Washington, D.C.: National Academy Press, 1984), p. 25.

²³Testimony of Erik D. Olson and Dr. Geral V. Poje, National Wildlife Federation, before the Subcommittee on Power Resources of the U.S. House Interior and Insular Affairs Committee, hearings on Reclamation States Groundwater Protection and Management Act of 1987, September 17, 1987, p. 131.

²⁴California Assembly Office of Research, *The Leaching Fields: A Nonpoint Threat to Groundwater,* 1985, p. iii.

²⁵Bob Barles and Jerry Kotas, "Pesticides and the Nation's Groundwater," *EPA Journal,* May 1987, pp. 42–43.

²⁶Marjorie Sun, "Groundwater Ills: Many Diagnoses, Few Remedies," *Science,* 20 June 1986, pp. 1490–1493.

²⁷Ibid.

²⁸U.S. EPA, "Drinking Water; Proposed Substitution of Contaminants and Proposed List of Additional Substances Which May Require Regulation Under the Safe Drinking Water Act," *52 Federal Register,* 8 July 1987, p. 25723.

²⁹R.H. Harris, *Implications of Cancer-Causing Substances in Mississippi River Waters* (Washington, D.C.: Environmental Defense Fund, November 1974); Talbot Page, et al., "Drinking Water and Cancer Mortality in Louisiana," *Science,* 193, No. 55 (1976), 55–57.

[30]Betty Dowty, "Halogenated Hydrocarbons in New Orleans Drinking Water and Blood Plasma," *Science*, 10 January 1975, pp. 75–77.

[31]S.W. Lagakos, B.J. Wessen, and M. Zelen, "An Analysis of Contaminated Well Water and Health Effects in Woburn, Massachusetts," *Journ. American Statistical Assoc.,* September 1986, pp. 583–596.

[32]Robert Hanley, "Women's Leukemia Rate Higher in 4 Jersey Towns," *The New York Times,* 13 December 1987, p. 64.

[33]California Department of Health Services, *Water Exposure and Pregnancy Outcomes, Summaries of Epidemiological Studies and Water Testing* (Berkeley, CA: Epidemiologcial Studies Section, 1988).

[34]National Academy of Sciences, *Drinking Water and Health, Vol. 7, Disinfectants and Disinfectant Byproducts,* Safe Drinking Water Committee, National Research Council (Washington, D.C.: National Academy Press, 1987).

[35]Herman F. Kraybill, "Carcinogenesis Induced by Trace Contaminants in Potable Water," *Bull. NY Acad. Med.,* 54, No. 4 (1978), 413–427.

4. The Limitations of Science and Regulation

[1]Mark E. Rushefsky, *Making CancerPolicy* (Albany: State University of New York Press, 1986).

[2]William D. Ruckelshaus, *Risk Analysis,* 4 (1984), 157–167.

[3]William D. Ruckelshaus, "Science Risk, and Public Policy," *Science,* 221, September 1983, 1026–1028.

[4]National Cancer Institute, *Human Health Considerations of Carcinogenic Organic Chemical Contaminants in Drinking Water,* Position Paper (Bethesda, MD: National Institute of Health, April 1978), p. 4.

[5]Richard Doll and Richard Peto, "The Causes of Cancer: Quantitative Estimates of Avoidable Risks of Cancer in the United States Today," *JNCI,* 66, No. 6 (1981), 1191–1308.

[6]OSHA, "Identification, Classification and Regulation of Potential Occupational Carcinogens," *45 Federal Register,* 22 January 1980, pp. 5002–5296.

[7]Comments of the Environmental Defense Fund, In: U.S. EPA, "National Interim Primary Drinking Water Regulations; Control of Trihalomethanes in Drinking Water," *44 Federal Register,* 29 November 1979, p. 68668.

5. Groundwater

[1]Jay H. Lehr, et. al., *A Manual of Laws, Regulations, and Institutions for Control of Groundwater Pollution,* Final Report, EPA-440/9-76-006 (Washington, D.C.: U.S. EPA, June 1976), p. I-6.

[2]Testimony of the U.S. Geological Survey before the Subcommittee on Toxic Substances and Environmental Oversight of the U.S. Senate Committee on Environment and Public Works, hearings on the SDWA Amendments of 1982, July 28, 1982, p. 584.

[3]Testimony of Philip Cohen, Chief Hydrologist, U.S. Geological Survey, before the Subcommittee on Toxic Substances and Environmental Oversight of the U.S. Senate Committee on Environment and Public Works, hearings on Groundwater Contamination and Protection, June 17, 1985, p. 17.

[4]J.H. Lehr, "How Much Groundwater Have We Really Polluted?," *Groundwater Monitoring Review,* Winter 1982, pp. 4–5.

[5]U.S. Office of Technology Assessment, *Protecting the Nation's Groundwater from Contamination,* OTA-O-233 (Washington, D.C.: U.S. OTA, October 1984), p. 21.

[6]Testimony of Doyle G. Fredrick, Associate Director, U.S. Geological Survey, before the Subcommittee on Water and Power of the U.S. Senate Committee on Energy and Natural Resources, hearings on Groundwater-Related Programs of the USGS and the EPA, June 5, 1987, p. 28.

[7]Ruth Patrick, Emily Ford, and John Quarles, *Groundwater Contamination in the United States,* 2nd ed. (Philadelphia, PA: University of Pennsylvania, 1987), p. 113.

[8]Testimony of Anthony D. Cortese, Commissioner of the Dept. of Env. Quality Engineering, sent to the Subcommittee on Environment, Energy and Natural Resources of the Committee on Government Operations, regarding Groundwater Contamination and the need for National Groundwater Strategy as exemplified in Massachusetts, July 24, 1980.

[9]James J. Westrick, Wayne Mello, and Robert F. Thomas, "The Groundwater Supply Survey," **JAWWA,** May 1984, pp. 52–59.

[10]U.S. EPA, **Report to Congress: Waste Disposal Practices and Their Effects on Ground-water** (Washington, D.C.: Office of Water Supply, January 1977).

[11]U.S. OTA, *Protecting the Nation's Groundwater From Contamination,* p. 1.

[12]Douglas M. Mackay, Paul V. Roberts, and John A. Cherry, "Transport of Organic Contaminants in Groundwater," **Environ. Sci. Technol.** 19, No. 5 (1985), 384–392.

[13]Ibid, p. 390.

[14]U.S. OTA, *Protecting the Nation's Groundwater from Contamination,* p. 22.

[15]Mackay, et. al, pp. 384–392.

[16]U.S. EPA, **Protection of Public Water Supplies from Groundwater Contamination,** Seminar Publication, EPA 625/4-85/016 (Cincinnati, OH: U.S. EPA, September 1985), p. 129.

[17]The Conservation Foundation, **Groundwater Protection** (Washington, D.C.: The Conservation Foundation, 1987), p. 60.

[18]Patrick, et. al., p. 55.

[19]David W. Miller, "Chemical Contamination of Groundwater," In: **Groundwater Quality,** C. Ward, W.Giger, P. McCarty eds. (New York: John Wiley & Sons, 1985), p. 45.

[20]Patrick, et. al., p. 56.

[21]Ibid.

[22]U.S. EPA, **Report to Congress.**

[23]Fletcher G. Driscoll, **Groundwater and Wells,** 2nd ed. (St. Paul: Johnson Division, 1986).

[24]U.S. EPA, **Protection of Public Water Supplies from Groundwater Contamination,** p. 110.

[25]Westrick, et. al., p. 57.

[26]Mackay, p. 387.

[27]Nancy K. Kim and Daniel W. Stone, **Organic Chemicals and Drinking Water** (Albany: New York State Department of Health, April 1981), p. 11.

[28]U.S. OTA, *Protecting the Nation's Groundwater From Contamination,* p. 43.

[29]U.S. EPA, "National Primary Drinking Water Regulations; Volatile Synthetic Organic Chemicals," **49 Federal Register,** 12 June 1984, p. 24335.

[30]Thomas Fusillo, et. al., "Distribution of Volatile Organic Compounds in a New Jersey Coastal Plain Aquifer System," **Groundwater,** May/June 1985, pp. 354–360.

[31]Gina Maranto, "The Creeping Poison Underground," **Discover,** February 1983, pp. 74–78.

[32]Ibid, p. 77.

[33]Julian Josephson "Restoration of Aquifers," **Environ. Sci. Tech.,** 17, No. 8 (1983), 350A.

6. Water Hardness & Cardiovascular Disease

[1]Henry A. Schroeder, "Relation Between Mortality From Cardiovascular Disease and Treated Water Supplies," **JAMA,** 172, No. 17 (1960), 1902–1908.

[2]National Academy of Sciences, **Drinking Water and Health,** Safe Drinking Water Committee, National Research Council (Washington, D.C.: National Academy of Sciences, 1977), p. 440.

[3]Ibid.

[4]Henry A. Schroeder and Luke A. Kraemer, "Cardiovascular Mortality, Municipal Water and Corrosion," **Arch. Environ. Health,** 28, June 1974, 310.

[5]Ronnie Levin, **Reducing Lead in Drinking Water: A Benefit Analysis,** Final Draft Report, EPA 230-09-86-019 (Washington, D.C.: Office of Policy, Planning and Evaluation, U.S. EPA, 1986), pp. IV-28.

[6]National Academy of Sciences, p. 441.

[7]Levin, pp. IV–29; National Academy of Sciences,. p 441.

[8]National Academy of Sciences, **Drinking Water and Health, Vol. 3,** Safe Drinking Water Committee, National Research Council (Washington, D.C.: National Academy Press, 1980), p. 22.

[9]K.R. Mahaffey and J. Rader, "Metabolic Interactions: Lead, Calcium and Iron," **Annals of the New York Academy of Sciences,** 355 (1980), 285–297.

[10]National Academy of Sciences, 1977, p. 442.

[11]Schroeder and Kraemer, p. 308.

[12]National Academy of Sciences, 1980, p. 23.

[13]National Academy of Sciences, 1977, p. 442.

7. Heavy Metal Pollutants

[1]H.A. Waldron, "Lead Poisoning in the Ancient World, **Medical History,** 17 (1973), 391–399.

[2]Margaret Engel, "City May Replace Lead Pipes," **The Washington Post,** 29 January 1987, p. 1.

[3]National Research Council, **Lead: Airbone Lead in Perspective** (Washington, D.C.: National Academy of Sciences, 1972).

[4]U.S. EPA, "National Primary Drinking Water Regulations, Proposed Rulemaking," **50 Federal Register,** 13 November 1985, p. 46968 .

[5]Marie Anne Urbanowicz, "The Uses of Lead Today," In: **Lead Toxicity,** Richard Lansdown and William Yule, eds. (Baltimore: The Johns Hopkins University Press, 1986), pp. 25–40.

[6]Waldron, p. 394.

[7]"Report of Commissioners, appointed by authority of the City Council to Examine the Sources from which a Supply of Pure Water may be obtained for the City of Boston," J.H. Eastburn, City Printer, Boston, 1845.

[8]William R. Nichols, "Poisoning by Lead Pipe Used for the Conveyance of Drinking Water," State Board of Health, Senate No. 50, January 1871, pp. 22–40.

[9]Wellington Donaldson, "The Action of Water on Service Pipes," **JAWWA,** 11, No. 5(1924), 649–662.

[10]Testimony of Senator Frank R. Lautenberg, before the U.S. Senate Committee on Environment and Public Works, hearings on the Lead Free Drinking Water Act, May 23, 1985, p. 3.

[11]U.S. EPA, "Drinking Water Regulations; Maximum Contaminant Level Goals and National Primary Drinking Water Regulations for Lead and Copper; Proposed Rule," **53 Federal Register,** 18 August 1988, p. 31519.

[12]Ibid, p. 31521.

[13]H.E. Hudson Jr., and F.W. Gikreas, "Health and Economic Aspects of Water Hardness and Corrosiveness," **JAWWA,** April 1976, pp. 201–204.

[14]U.S. EPA, **53 Federal Register,** p. 31526.

[15]Walter Sullivan, "High Acidity in a Jersey Aquifer Is Found to Increase Lead Hazard," **The New York Times,** 20 May 1987, p. D31.

[16]A. Britton and W.N. Richards, "Factors Influencing Plumbosolvency in Scotland," **Journal Inst. Water Eng. Sci.,** 35, No. 4 (1981), 349–364.

[17]Testimony of Norman Murrell, Senior Vice President, H2M, before the U.S. Senate Committee on Environment and Public Works, hearings on the Lead Free Drinking Water Act, May 23, 1985, p. 16.

[18]Ibid.

[19]Ibid, Testimony of Senator Dave Durenberger, p. 7.

[20]U.S. EPA, **53 Federal Register,** p. 31521.

[21]Ibid, p. 31526.

[22]"Lead Pollution Declined in '85," **The New York Times,** 18 April 1987, p.13.

[23]T.G. Lovering, "Lead in the Environment," **U.S. Geological Survey Paper #957** (Washington, D.C.: U.S. GPO, 1976).

[24]H.J. Abrams, "The Water Supply of Rome," *JAWWA*, 67 (1975), 663-668.

[25]S.K. Hall, "Lead Pollution and Poisoning," *Environ. Sci. Technol.*, 6 (1972) p. 31.

[26]Marjorie Smith, "Lead in History," In: *Lead Toxicity*, Richard Lansdown and William Yule, eds. (Baltimore: The Johns Hopkins University Press, 1986), pp. 7–24.

[27]Testimony of Senator Bill Bradley, before the U.S. Senate Committee on Environment and Public Works, hearings on The Lead Free Drinking Water Act, May 23, 1985, p. 5.

[28]William N. Richards and Michael R. Moore, "Lead Hazard in Scottish Water Systems," *JAWWA*, August 1984, pp. 60–67.

[29]Janet Hunter, "The Distribution of Lead," In: *Lead Toxicity History and Environmental Impact*, Richard Lansdown and William Yule, eds. (Baltimore: The Johns Hopkins University Press, 1986), pp. 96–130.

[30]Roy M. Harrison and Duncan P.H. Laxen, "Physiochemical Speciation of Lead in Drinking Water," *Nature*, 286, No. 21 (1980), 791–793.

[31]Michael R. Moore, William N. Richards, and John G. Sherlock, "Successful Abatement of Lead Exposure from Water Supplies in the West of Scotland," *Environmental Research*, 38 (1985), 67–76.

[32]Ronnie Levin, *Reducing Lead in Drinking Water: A Benefit Analysis*, Draft Final Report, EPA-230-09-86-019 (Washington, D.C.: Office of Policy, Planning & Evaluation, U.S. EPA, December 1986).

[33]A.D. Beattie, et. al, "Role of Chronic Low-Level Lead Exposure in the Etiology of Mental Retardation," *The Lancet*, 15 March 1975, pp. 569–592.

[34]Ibid.

[35]David Bellinger, et. al, "Longitudinal Analysis of Prenatal and Postnatal Lead Exposure and Early Cognitive Development," *New England Journal of Medicine*, 316, No. 17 (1987), 1037–1043.

[36]Testimony of Norman Murrell, p. 17.

[37]U.S. EPA, *Lead and Your Drinking Water*, OPA-87-006 (Washington, D.C.: Office of Water, April 1987).

[38]Ramon Lee, William Becker, David Collins, "Lead at the Tap: Sources and Control," *JAWWA*, July 1989, pp. 52–62.

[39]Levin, p. I-20.

[40]U.S. EPA, *Lead and Your Drinking Water.*

[41]Ibid.

[42]National Academy of Sciences, *Drinking Water and Health*, Vol. 4, Safe Drinking Water Committee, National Research Council (Washington, D.C.: National Academy Press, 1981), p. III-13.

[43]"Regulating Organics," interview with Michael Cook, director, U.S. EPA, Office of Drinking Water, *JAWWA*, January 1987, pp. 10–23.

[44]U.S. EPA, *53 Federal Register*, p. 31532.

[45]David A. Grantham and John F. Jones, "Arsenic Contamination of Water Wells in Nova Scotia," *JAWWA*, December 1977, pp. 653–703.

[46]Edward J. Feinglass, "Arsenic Intoxication from Well Water in the United States," *New England Journal of Medicine*, 288, No. 16 (1973), 828–830.

[47]Lowenbach and Schlesinger Associates, *Arsenic: A Preliminary Materials Balance*, EPA 560/6-79-005 (Washington, D.C.: Office of Toxic Substances, U.S. EPA, March 1979).

[48]U.S. EPA, *50 Federal Register*, p. 46959–60.

[49]U.S. EPA, *50 Federal Register*, p. 46960.

[50]National Academy of Sciences, *Drinking Water and Health*, Safe Drinking Water Committee, National Research Council (Washington, D.C.: National Academy of Sciences, 1977), p. 319.

[51]U.S. EPA, *50 Federal Register*, p. 46960.

[52]Robert Zaldivar, "Arsenic Contamination of Drinking Water and Foodstuffs Causing Endemic Chronic Poisoning," *Beitr. Path. Bd.*, 151 (1974), 384–400.

[53]Michael Weisskoff, "EPA Panel Shifts Course on Pollutant," *The Washington Post,* 5 October 1987, p. A13.

[54]U.S. EPA, *50 Federal Register,* p. 46960.

[55]Statement of Ben T. Bartlett, City Manager, City of Fallon, submitted to the EPA to be considered in the Proposed Rulemaking for an RMCL for Arsenic in Drinking Water.

[56]Ibid.

[57]National Academy of Sciences, *Arsenic,* Committe on Medical and Biological Effects of Environmental Pollutants, National Research Council (Washington, D.C.: National Academy of Sciences, 1977).

[58]U.S. EPA, *50 Federal Register,* p. 46964.

[59]William D. Rowe, et. al., "The Nature and Pollution Sources of Cadmium," In: *Evaluation Methods for Environmental Standards* (Boca Raton, FL: CRC Press, 1983), pp. 53–55.

[60]Rowe, p. 55.

[61]John M. Lucas, *Cadmium, Mineral Commodity Profiles,* Bureau of Mines (Washington, D.C.: U.S. Dept. of the Interior, August 1979).

[62]C. Elinder and K. Jellstrom, "Total Cadmium Concentrations in Samples of Human Kidney Cortex from the 19th Century," *Ambio,* 6 (1977), 270.

[63]U.S. EPA, *50 Federal Register,* p. 46965; James Ryan, Herbert Pahren, and James Lucas, "Controlling Cadmium in the Human Food Chain: A Review and Rationale Based on Health Effects," *Environmental Research,* 28 (1982), 251–302; R. Lauweys and P. DeWals, "Environmental Pollution by Cadmium and Mortality from Renal Diseases," *The Lancet,* 14 February 1981, p. 383.

[64]U.S. EPA, *50 Federal Register,* p. 46965.

[65]"Science Watch: Cadmium and Alcoholism," *The New York Times,* 4 August 1987, p. 66.

[66]U.S. EPA, *50 Federal Register,* p. 46968.

[67]National Academy of Sciences, 1977, p. 308.

[68]W.C. Butterman, *Mineral Commodity Profiles: Copper,* Bureau of Mines (Washington, D.C.: U.S. Dept. of Interior, 1983).

[69]Kenneth C. Spitalny, et. al., "Drinking-Water- Induced Copper Intoxification in a Vermont Family," *Pediatrics,* 74, No. 6 (1984), 1103–1106.

[70]U.S. EPA, *50 Federal Register,* p. 46968.

[71]U.S. EPA, *Ambient Water Quality Criteria for Chromium,* EPA 440/5-80-035 (Washington, D.C.: Criteria and Standards Division, October 1985), p. C-2.

[72]Fredrick N. Robertson, "Hexavalent Chromium in the Groundwater in Paradise Valley, Arizona," *Groundwater,* 13, No. 6 (1975), 516–527.

[73]Ibid.

[74]World Health Organization, *Guidelines for Drinking Water Quality, Vol. 2, Health Criteria and Other Supporting Information* (Geneva: WHO, 1984), pp. 91–96.

[75]U.S. EPA, *50 Federal Register,* p. 46966.

[76]World Health Organization, p. 95.

[77]U.S. EPA, *Ambient Water Quality Criteria for Chromium,* p. C-3.

[78]P.A. D'Itri and F.M. D'Itra, *Mercury Contamination: A Human Tragedy* (New York: Wiley-Intescience, 1977).

[79]K. Irukayama, *The Pollution of Minamata Bay and Minamata Disease,* Proc. Third International Conf. of Advances in Water Pollution Research, Vol. 3 (Washington, D.C.: Water Pollution Control Federation, 1967), p. 153.

[80]U.S. EPA, "National Revised Drinking Water Regulations; Advanced Notice of Rulemaking," *48 Federal Register,* 5 October 1983, p. 45513.

[81]Thomas R. Stolzenburg, Robert Stanforth and David Nichols, "Potential Health Effects of Mercury in Water Supply Wells," *JAWWA,* January 1986, pp. 45–48.

[82]D'Itri and D'Itri, p. 146.

8. Biological Contamination

[1]National Academy of Sciences, *Drinking Water and Health,* Safe Drinking Water Committee, National Research Council (Washington D.C.: National Academy of Sciences, 1977), p. 1.

[2]Edwin E. Geldreich, Martin J. Allen, and Raymond H. Taylor, "Interferences to Coliform Detection in Potable Water Supplies," In: *Evaluation of Microbiology Standards for Drinking Water,* Charles W. Hendricks, ed., EPA-570/9-78-OOC (Washington, D.C.: Office of Drinking Water, U.S. EPA, 1978), p. iii.

[3]Kenneth P. Cantor, "Epidemiological Evidence of Carcinogenicity of Chlorinated Organics in Drinking Water," *Environ. Health Persp.,* 46 (1982), 187–195.

[4]G.I. Barrow, "Bacterial Indicators and Standards of Water Quality in Britain," *Bacterial Indicators/Health Hazards Associated with Water,* A.W. Hoadley and B.J. Dutka, eds., ASTM STP 635, American Society for Testing and Materials (1977), p. 296.

[5]Ibid, p. 297.

[6]Centers for Disease Control, *Water-Related Disease Outbreaks, Annual Summary for 1984* (Atlanta, GA: CDC, 1985).

[7]U.S. EPA, "Drinking Water; National Primary Drinking Water Regulations; Total Coliforms (Including Fecal Coliforms and E. Coli); Final Rule," *54 Federal Register,* 29 June 1989, p. 27547.

[8]Ibid.

[9]Ramon J. Seidler and Thomas M. Evans, *Persistence and Detection of Coliforms in Turbid Finished Drinking Water,* Project Summary (Cincinnati, OH: U.S. EPA, 1982).

[10]Gunther F. Craun, "A Summary of Waterborne Illness Transmitted through Contaminated Groundwater," *J. Environ. Health,* 48, No. 3 (1985), 122–127.

[11]Geldreich, et. al., "Interferences to Coliform Detection in Potable Water Supplies," p. 14.

[12]Edwin Geldreich, et. al., "The Necessity of Controlling Bacterial Populations in Potable Waters: Community Water Supply," *JAWWA,* 64, No. 9 (1972), 598.

[13]Craun, p. 123.

[14]Ibid, p. 123.

[15]Stuart R. Crane, and James A. Moore, "Bacterial Pollution of Groundwater: A Review," *Water, Air, and Soil Pollution,* 22 (1984), 77.

[16]Crane and Moore, p. 67–83.

[17]Odette Batik, et. al., "Routine Coliform Monitoring and Waterborne Disease Outbreaks," *J. Environ. Health,* March/April 1983, p. 227.

[18]U.S. EPA, *National Interim Primary Drinking Water Regulations,* EPA 570/9-76-003 (Washington, D.C.: U.S. Government Printing Office, 1976).

[19]Mark W. LeChevalier and Gordon A. McFeters, "Recent Advances in Coliform Methodology for Water Analysis," *J. Environ. Health,* 47, No. 1 (1984), 6.

[20]U.S. EPA, *54 Federal Register,* 1989, p. 27549.

[21]R.R. Colwell, et. al., "Public Health Considerations of the Microbiology of 'Potable' Water," In: *Evaluation of Microbiology Standards for Drinking Water,* Charles W. Hendricks, ed., EPA 570/9-78-OOC (Washington, D.C.: Office of Drinking Water, 1978), p. 65.

[22]Batik, et. al., p. 229.

[23]Leland J. Mcabe, "Chlorine Residual Substitution-Rationale," In: *Evaluation of the Microbiology Standards for Drinking Water,* Charles W. Hendricks, ed., EPA 570/9-78-OOC (Washington, D.C.: Office of Drinking Water, 1978), p. 57.

[24]Batik, et. al., p. 227.

[25]Gunther F. Craun, "Impact of the Coliform Standard on the Transmission of Disease," In: *Evaluation of the Microbiology Standards for Drinking Water,* Charles W. Hendricks, ed., EPA 570/9-78-OOC (Washington, D.C.: Office of Drinking Water, 1978), p. 28.

[26]U.S. EPA, "Drinking Water; National Primary Drinking Water Regulations; Filtration, Disinfection; Turbidity, Giardia Lamblia, Viruses, Legionella, and Heterotrophic Bacteria; Final Rule," *54 Federal Register,* 29 June 1989, p. 27489.

[27]Am. Pub. Health Assoc., Am. Water Works Assoc., and Water Pol. Control Fed., *Standard Methods for the Examination of Water and Wastewater,* 16th ed. (Washington, D.C.: APHA, 1985), p. 860.

[28]U.S. EPA, "Drinking Water; National Primary Drinking Water Regulations; Total Coliforms; Proposed Rule," **52 Federal Register,** 3 November 1987, pp. 42224–42245.

[29]U.S. EPA, "National Primary Drinking Water Regulations; Synthetic Organic Chemicals, Inorganic Chemicals and Microorganisms; Proposed Rule," **50 Federal Register,** 13 November 1985, p. 46956.

[30]Edwin E. Geldreich, **Handbook for Evaluating Water Bacteriological Laboratories,** 2nd ed., EPA-670/9-75-006 (Cincinnati, OH: U.S. EPA, August 1975), p. 19.

[31]Geldreich, **Handbook for Evaluating Water Bacteriological Laboratories,** p. 20.

[32]Geldreich, et. al., "The Necessity of Controlling Bacterial Populations in Potable Waters," p. 600.

[33]Am. Pub. Health Assoc., et. al., p. 860.

[34]Ibid, p.859.

[35]Audrey E. McDaniels, "Holding Effects on Coliform Enumeration in Drinking Water Samples," **Appl. Env. Micro.,** October 1985, pp. 755–762.

[36]Geldreich, et. al., "Interferrences to Coliform Detection in Potable Water Supplies," p. 13; **Am. Pub. Health Assoc., et. al., p. 133.**

[37]**U.S. EPA, 50 Federal Register,** p. 46952.

[38]Geldreich, "Membrane Filter Techniques...," p. 45.

[39]James W. Smith, "Giardiasis," **Ann. Rev. Med.,** 31 (1980), 373.

[40]"Waterborne Giardiasis, Where and Why," **JAWWA,** January 1986, pp. 85–86.

[41]Ibid.

[42]Jerry E. Ongerth, "Giardia Cyst Concentrations in River Water," **JAWWA,** September 1989, pp. 81–86.

[43]Batik, et. al., p. 230.

[44]"Waterborne Giardiasis, Where and Why," p. 85.

[45]David A. Buckner, et. al., "Intestinal Parasites in Los Angeles, California," **Am. J. Med. Technol.,** 45, No. 12 (1979), 1020.

[46]Centers for Disease Control, "Bacterial Diseases," **Veterinary Public Health Notes** (Atlanta, GA: U.S. Public Health Service, January 1979).

[47]Smith, pp. 376–377.

[48]"Waterborne Giardiasis, Where and Why," p. 85.

[49]G.F. Craun, "Waterborne Outbreaks of Giardiasis," In: **Waterborne Transmission of Giardiasis, Proc. of a Symposium, September 18–20,** W. Jakubowski, and J.C. Hoff, eds., EPA-600/9-79-001 (Cincinnati, OH: U.S. EPA, June 1979), p. 140.

[50]"Waterborne Giardiasis, Where and Why," p. 85.; Centers for Disease Control, **Water-Related Disease Outbreaks, Annual Summary 1984,** Issued November 1985, p. 2.

[51]Peter K. Show, et. al., "A Communitywide Outbreak of Giardiasis with Evidence of Transmission by a Municipal Water Supply," **Ann. Int. Med.,** 87 (1977), 426–432.

[52]Bruce G. Weniger, et. al., "An Outbreak of Waterborne Giardiasis Associated with Heavy Water Runoff Due to Warm Weather and Volcanic Ashfall," **Am. J. Pub. Health,** 73, August 1983, 868–872.

[53]U.S. EPA, **54 Federal Register,** 1989, p. 27489.

[54]Lucy Harter, et. al., "Giardia Prevalence Among 1 to 3 Year-Old Children in Two Washington State Counties," **Am. J. Pub. Health,** 72, No. 4 (1982), 386–388.

[55]Floyd Frost, et. al., **Giardiasis in Washington State** (Research Triangle Park, N.C.: U.S. EPA, December 1981), p. 3.

[56]J.C. Kirner, et. al., "A Waterborne Outbreak of Giardiasis in Camas, Washington," **JAWWA,** January 1978, pp. 35–40; Shun Dar Lin, "Giardia Lamblia and Water Supply," **JAWWA,** February 1985, pp. 40–47.

[57]Edwin C. Lippy, "Tracing a Giardiasis Outbreak in Berlin, New Hampshire," **JAWWA,** September 1978, p. 519.

[58]Craun, "Waterborne Outbreaks of Giardiasis," p. 144.

[59]Ibid, p. 143.

[60]U.S. EPA, "Drinking Water; National Primary Drinking Water Regulations; Filtration, Disinfection; Turbidity, Giardia Lamblia, Viruses, Legionella, and Heterotrophic Bacteria; Final Rule," **54 Federal Register,** 29 June 1989, p. 27494.

[61]U.S. EPA, "National Primary Drinking Water Regulations; Filtration and Disinfection; Turbidity, Giardia Lamblia, Viruses, Legionella, and Heterotrophic Bacteria; Proposed Rule," **52 Federal Register,** 3 November 1987, p. 42195.

[62]Kelly P. Lange, et. al., "Diatomaceous Earth Filtration of Giardia Cysts and Other Substances," **JAWWA,** January 1986, pp. 77–83.

[63]Samuel Strotynski and Thomas A. Reamon, "Giardia as Related to Water," New York State Dept. of Health, Div. of Sanitary Eng., Bureau of Pub. Water Supply, undated, p. 4.

[64]Dean O. Cliver, "Virus Detection," In: *Evaluation of the Microbiology Standards for Drinking Water,* Charles W. Hendrick, ed., EPA 570/9-78-OOC (Washington, D.C.: Office of Drinking Water, 1978), p. 95.

[65]John T. Cookson, Jr., "Virus and Water Supply," **JAWWA,** December 1974, p. 709.

[66]Bruce H. Keswick, et. al., "Inactivation of Norwalk Virus in Drinking Water by Chlorine," **App. Environ. Micro.,** 50, No. 2 (1985), 261–264.

[67]U.S. EPA, **52 Federal Register,** 1987, p. 42195.

[68]Cookson, p. 707.

[69]Ibid, p. 710.

[70]U.S. EPA, **52 Federal Register,** 1987, p. 42195.

[71]Ibid, p. 42195.

[72]World Health Organization, **Human Viruses in Water, Wastewater and Soil,** Technical Report Series 639 (Geneva: WHO, 1979).

[73]"Detection of Viral Contamination in Groundwater and Its Relationship to the Presence of Coliform Bacteria," Water Quality Technology Conference Abstracts, **JAWWA,** 79, No. 9 (1987), 50.

[74]WHO, p. 31.

[75]Bruce Keswick and Charles Gerba, "Viruses in Groundwater," **Environ. Sci. Tech.,** 14, No. 11 (1980), 1290–1297.

[76]Ibid, p. 1295.

[77]Gabriel Bitton, et. al., "Viruses in Drinking Water," **Environ. Sci. Tech.,** 20, No. 3 (1986), 216–222.

[78]Charles P. Gerba, et. al., *Viruses Removal During Conventional Drinking Water Treatment,* Project Summary (Research Triangle Park, N.C.: U.S. EPA, September 1985).

[79]Bitton, et. al., p. 219.

[80]Ibid.

[81]WHO, p. 33.

[82]Linden E. Witherell, et. al., *Investigation of Legionella Pneumophila in Drinking Water,* Project Summary, EPA/600/51-85/019 (Research Triangle Park, N.C.: U.S. EPA, September 1985).

[83]Victoria Churchville, "Legionnaire's Disease Probed in 2 Maryland Deaths," **The Washington Post,** 3 June 1986, p. 1.

[84]Lori B. Miller, "After A Death, Workers Tested for a Sickness," **New York Times,** 10 July 1987, p. B3.

[85]"New York Times Workers Had Legionnaire's Disease," **The Wall Street Journal,** 31 July 1985, p. 25.

[86]Janet E. Stout, et. al., "Ecology of Legionella Pneumophila Within Water Distribution Systems," **Appl. Env. Micro.,** January 1985, pp. 221–228.

[87]Witherell, et. al., p. 2.

[88]Carol A. Ciesielski, "Role of Stagnation and Obstruction of Water Flow in Isolation of Legionella Pneumophila from Hospital Plumbing," **Appl. Env. Micro.,** November 1984, pp. 984–987.

[89]John M. Kuchta, et. al., "Susceptibility of Legionella Pneumophila to Chlorine in Tap Water," *Appl. Env. Micro.*, November 1983, pp. 1134–1139.

[90]Janet Stout, et. al., "Ubiquitousness of Legionella Pneumophila in the Water Supply of a Hospital with Endemic Legionnaire's Disease," *New Eng. J. Med.*, 25 February 1982, pp. 466–468.

[91]Gary E. Bollin, "Aerosols Containing Legionella Pneumophila Generated by Shower Heads and Hot-Water Faucets," *Appl. Env. Micro.*, November 1985, pp. 1128–1131.

[92]Stout, et. al., "Ecology of Legionella Pneumophila...," p. 221.

[93]U.S. EPA, *52 Federal Register,* p. 42181.

[94]Kuchta, et. al., p. 1138.

[95]Stout, et. al., "Ubiquitousness of Legionella Peneumophila..." p. 467.

[96]Stefi Weisburd, "Hunting for Legionnaire's Bacteria," *Science News,* 12 September 1987, p. 169.

[97]R.M. Wadowsky, et. al., "Hot Water Systems as Sources of Legionella Pneumophila in Hospital and Non Hospital Plumbing Fixtures," *App. Env. Micro.*, 43, No. 5 (1982), 1104–1110.

[98]Stout, et. al., "Ecology of Legionella Pneumophila," p. 227.

[99]Ibid.

[100]Ciesielski, et. al., p. 984.

[101]Lance Voss, et. al., "Legionella Contamination of a Preoperational Treatment Plant," *JAWWA,* January 1986, pp. 70–75.

[102]Robert B. Yee and Robert M. Wadowsky, "Multiplication of Legionella Pneumophila in Unsterilized Tap Water," *Appl. Env. Micro.*, June 1982, pp. 1330–1334.

[103]Witherall, et. al.

[104]Jane Stout, et. al., "Legionnaire's Disease Acquired Within the Homes of Two Patients — Link to the Home Water Supply," *JAMA,* 257, No. 9 (1987), 1215–1217.

[105]Kutchta, et. al., p. 1138.

[106]Witherall, et. al., p. 3.

[107]David W. Fraser, "Potable Water as a Source for Legionellosis," *Environ. Health Persp.,* 62 (1985), 337–441.

[108]Ibid, p. 339.

9. Radioactivity

[1]U.S. EPA, "National Primary Drinking Water Regulations; Advanced Notice of Primary Regulations; Radionuclides," *51 Federal Register,* 30 September 1986, p. 34839.

[2]Ibid.

[3]J.D. Lowry and S. Lowry, "Radionuclides in Drinking Water," *JAWWA,* July 1988, pp. 50-64.

[4]C.T. Hess et. al., "The Occurrence of Radioactivity in Public Water Supplies in the United States," *Health Physics,* 48, No. 5 (1985), 553–586.

[5]E. Marco Aieta, et. al., "Radionuclides in Drinking Water: An Overview," *JAWWA,* April 1987, pp. 144–152.

[6]Charles D. Strain, and James E. Watson, "An Evaluation of Radium-226 and Radon-222 Concentrations in Ground and Surface Water Near a Phosphate Mining and Manufacturing Facility," *Health Physics,* 37 (1979), 779–783.

[7]C. Richard Cothern, and William L. Lappenbusch, "Occurrence of Uranium in Drinking Water in the U.S.," *Health Physics* 45, No. 1 (1983), 89–99.

[8]W.L. Lappenbusch and Moskowitz, U.S. EPA Health Advisory Program, AMA Smposium on Drinking Water and Human Health, 1983.

[9]Henry F. Lucas, "Radium-226 and Radium-228 in Water Supplies," *JAWWA,* Sept. 1985, pp. 57–67.

[10]Hess, et. al., 1985, p. 566.

[11]J.D. Lowry and S.B. Lowry, p. 57.

[12]Ibid.

[13]J.D. Lowry and S.B. Lowry, p. 53.

[14]Jacqueline Michel and C. Richard Cothern, "Predicting the Occurrence of Ra-228 in Groundwater," **Health Physics,** 51, No. 6 (1986), 715–721.

[15]Bernard L. Cohen, "A National Survey of Rn222 in U.S. Homes and Correlating Factors," **Health Physics,** 51, No.2 (1986), 175–183.

[16]C.T. Hess, C.V. Weiffenbach, and S.A. Norton, "Variations of Airborne and Waterborne Rn-222 in Houses in Maine," **Environment International,** 8 (1982), 59–66.

[17]U.S. EPA, **51 Federal Register,** p. 34842.

[18]Jerry D. Lowry, et. al., "Extreme Levels of Radon-222 and Uranium in a Private Water Supply," In: **Radon, Radium, and Other Radioactivity in Groundwater - Hydrogeologic Impact and Application to Indoor Airborne Contamination,** Barbara Graves, ed. (Chelsea: Lewis Publisher, 1987), pp. 363–375.

[19]Howard M. Prichard, "The Transfer of Radon From Domestic Water to Indoor Air," **JAWWA,** April 1987, pp. 159–161.

[20]Fredrick W. Pontius, "Complying with the New Drinking Water Quality Regulations," **JAWWA,** February 1990, pp. 33–52.

[21]William L. Lappenbusch and C. Richard Cothern, "Regulatory Development of the Interim and Revised Regulations for Radioactivity in Drinking Water — Past and Present Issues and Problems," **Health Physics,** 48, No. 5 (1985), 535–551.

[22]U.S. EPA, **51 Federal Register,** p. 34841.

[23]Eleanor Charles, "The Radon Risk Brings a Surge in Testing," **The New York Times,** 12 July 1987, p. R12.

[24]J.D. Lowry and S.B. Lowry, p. 57.

[25]Hess, et. al., p. 568.

[26]Allan Richards, "Disaster at Church Rock: The Untold Story," **Sierra,** November/December 1980, p. 30.

[27]Hess, et. al., p. 572.

[28]Ibid.

[29]**Ionizing Radiation Exposure of the Population of the United States,** NCRP Report No. 93 (Bethesda, MD: National Council on Radiation Protection and Measurements, 1987), p.55.

[30]U.S. EPA, **Radioactivity in Drinking Water,** EPA 570/9-81-002 (Washington, D.C.: Health Effects Branch Criteria and Standards Division, January 1981), p. 29.

[31]U.S. EPA, **51 Federal Register,** pp. 34842–34843.

[32]William L. Lappenbusch, **Contaminated Drinking Water and Your Health** (Alexandria, VA: Lappenbusch Environmental Health, Inc., 1986), p. 133.

[33]Ibid.

[34]U.S. EPA, **A Citizen's Guide to Radon,** OPA-86-004 (Washington, D.C.: Office of Air and Radiation, August 1986).

[35]Maine Department of Human Services,**Radon in Water and Air: Health Risks and Control Measures** (Orono, ME: Land and Water Resources Center, Univ. of Maine, Orono, February 1983), pp. 1–12.

[36]C. Richard Cothern, "Estimating the Health Risks of Radon in Drinking Water," **JAWWA,** April 1987, pp. 153–159.

[37]J.D. Lowry and S.B. Lowry, p. 53.

[38]Ibid.

[39]Ibid.

[40]U.S. EPA, **51 Federal Register,** p. 34840.

[41]Ibid, p. 34841.

[42]U.S. GAO, **Nuclear Energy: Environmental Issues at DOE's Nuclear Defense Facilities,** GAO/RCED-86-192 (Washington, D.C.: U.S. GAO, September 1986).

[43]Ibid.

[44]Ibid.

⁴⁵C.R. Olsen, et. al., "Reactor-Released Radionuclides in Susquehanna River Sediments," *Nature,* 294 (1981), 242.

⁴⁶Statement of Thomas A. Luken, Chairman, House Subcommittee on Transporation, Tourism, and Hazardous Materials, hearings to Investigate DOE Actions to Circumvent and Undermine Efforts to Reduce Pollution at Fernald, OH Plant, October 14, 1988; "Consultant Puts Fernald Leaks at 1 Million Pounds," *The Cleveland Plain Dealer,* 3 May 1989, p. B3.

⁴⁷"Miami Aquifer Contaminated Near Fernald," *The Cleveland Plain Dealer,* 4 January 1990, p. B1; "Water at Fernald Tests Higher in Uranium Content," *The Cleveland Plain Dealer,* 9 February 1990, p. B2; "Poisoned Wells Found Near Fernald," *The Cleveland Plain Dealer,* 27 January 1990, p. B2.

⁴⁸Statement of Charles A. Bowster, Comptroller General of the United States, before the U.S. House Committee on the Budget, *Dealing with Enormous Problems in the Nuclear Weapons Complex,* GAO/T-RCED-89-6, February 8, 1989.

⁴⁹Carol Polsgrove, "In Hot Water: Uranium Mining and Water Pollution," *Sierra,* November/December 1980, pp. 28–31.

⁵⁰U.S. General Accounting Office, *Nuclear Waste, Problems Associated with DOE's Inactive Waste Sites,* GAO/RCED-88-169 (Washington, D.C.: U.S. GAO, August 1988).

⁵¹Lappenbusch, *Contaminated Drinking Water and Your Health,* p. 149.

⁵²C. Richard Cothern and William L. Lappenbusch, "Compliance Data For the Occurrence of Radium and Gross Alpha-Particle Activity in Drinking Water Supplies in the United States," *Health Physics,* 46, No. 3 (1984), 503–510.

⁵³Lappenbusch, *Contaminated Drinking Water and Your Health,* p. 141.

⁵⁴Lucas, pp. 57–67.

⁵⁵Hess, et. al., 1985, p. 563.

⁵⁶William Mills, William H. Ellett, and Robert E. Sullivan, "Monitoring for Ra-228 in Water Supplies," *Health Physics,* 39 (December), 1003.

⁵⁷U.S. EPA, *51 Federal Register,* p. 34841.

⁵⁸"Higher Radon Levels in Summer Cast Doubt on Reliability of Tests," *The New York Times,* 23 May 1989, p.C4.

⁵⁹Maine Dept. of Human Services, *Radon in Water and Air;* T.R. Horton, *Nationwide Occurrence of Radon and Other Natural Radioactivity in Public Water Supplies,* EPA-520/5-85-008 (Washington, D.C.: U.S. EPA, 1985).

⁶⁰U.S. EPA, *A Citizen's Guide to Radon.*

⁶¹"EPA Finds Radon Problems in 10-State Survey," *Environmental News* (Washington, D.C.: U.S. EPA Office of Public Affairs, August 1987).

⁶²U.S. General Accounting Office, *Air Pollution, Hazards of Indoor Radon Could Pose a National Health Problem,* GAO/RCED-86-170 (Washington, D.C.: U.S. GAO, June 1986).

⁶³Maine Dept. of Human Services, *Radon in Water and Air,* p. 6.

⁶⁴Herman L. Krieger, and Earl L. Whittaker, *Prescribed Procedures for Measurement of Radioactivity in Drinking Water,* EPA-600/4-80-032 (Cincinnati, OH: Env. Monitoring and Support Lab, U.S. EPA, August, 1980).

⁶⁵U.S. EPA, *Removal of Radon From Household Water* (Washington, D.C.: Office of Research and Development, U.S. EPA, September 1987).

⁶⁶Dennis Clifford, et. al., "Evaluating Various Adsorbents and Membranes for Removing Radium from Groundwater," *JAWWA,* July 1988, pp. 94–104.

⁶⁷J.D. Lowry and S.B. Lowry, p. 59.

⁶⁸Thomas Sorg, "Methods for Removing Uranium From Drinking Water," *JAWWA,* July 1988, pp. 105–111.

⁶⁹J.D. Lowry and S.B. Lowry, p. 62.

⁷⁰Jerry D. Lowry, Lowry Engineers, Inc., Thorndike, Maine, personal communication, June 5, 1987.

⁷¹Jerry D. Lowry, et. al., "Point-of-Entry Removal of Radon From Drinking Water," *JAWWA,* April 1987, p. 162–169.

[72]Ibid, p. 167.

[73]U.S. EPA, *Removal of Radon from Household Water.*

[74]U.S. EPA, *51 Federal Register,* p. 34848.

[75]Lowry, et. al., "Point-of-Entry Removal of Radon From Drinking Water," p. 167.

[76]Jerry D. Lowry, et. al., "Extreme Levels of Radon-222 and Uranium in a Private Water Supply," p. 365.

[77]J.D. Lowry and S.B. Lowry, p. 61.

[78]U.S. EPA, *51 Federal Register,* p. 34848.

10. Nitrates

[1]U.S. EPA, "National Primary Drinking Water Regulations; Proposed Rulemaking," **50** *Federal Register,* 13 November 1985, p. 46972.

[2]Harry Vroomen, *Fertilizer Use and Price Statistics 1960–85,* Economic Research Service, USDA, Statistical Bulletin No. 750, 1986.

[3]Council for Agricultural Science and Technology, *The Double-Edged Sword of Nitrogen Fertilizer* (Ames, IA: CAST, July 1983).

[4]"Fertilizer Companies Risk Fraud Charges," *Center for Rural Affairs Newsletter,* December 1987, p. 1–4.

[5]J.R. Gormly, and R.F. Spaulding, "Sources and Concentrations of Nitrate-Nitrogen Groundwater on the Platte Region, Nebraska," *Groundwater,* 17, No.3 (1979), 291–301.

[6]P.G. Saffigna and D.R. Keeney, "Nitrate and Chloride in Groundwater Under Irrigated Agriculture in Central Wisconsin," *Groundwater,* 15, No. 2 (1977), 170–177.

[7]E. Dickey, et.al., "Nitrate Levels and Possible Sources in Shallow Wells," In: *Proceedings: Second Allerton Conference on Environmental Quality and Agriculture,* University of Illinois at Urbana, Special Publication, No. 26 (1971), pp. 40–44.

[8]Kolby T. Crabtree, *Nitrate and Nitrite Variation in Groundwater,* Technical Bulletin No. 58 (Madison, Wisconsin: Dept. of Natural Resources, 1972), pp. 1–24.

[9]Marty Strange and Liz Krupicka, "Farming and Cancer," In: *It's Not All Sunshine and Fresh Air,* Center for Rural Affairs, April 1984, pp. 41–53.

[10]Graham Walton, "Survey of Literature Relating to Infant Methemoglobinemia Due to Nitrate-Contaminated Water," *Am. J. Pub. Health,* August 1951, pp. 986–995.

[11]National Academy of Sciences, *Drinking Water and Health,* Safe Drinking Water Committee, National Research Council (Washington, D.C.: National Academy of Sciences, 1977), p. 424.

[12]Council for Agricultural Science and Technology, p. 3.

[13]H.M. Bosch, et. al, "Methemoglobinemia and Minnesota Well Supplies," **JAWWA,** 42 (1950), 161–170.

[14]E. Hegesh and J. Shiloah, "Blood Nitrates and Infantile Methemoglobinemia," *Clinica Chimica Acta,* 125 (1982), 107–115.

[15]M. Cornblath and A.F. Hartmann, "Methemoglobinemia in Young Infants," *J. of Pediatrics,* 33 (1948), 421–425.

[16]Lois Ann Shearer, et.al., "Methemoglobin Levels in Infants in an Area with High Nitrate Water Supply," *Am. J. Pub. Health,* 62, No. 9 (1972), 1174–1180.

[17]Kenneth Cantor and Aaron Blair, *Agricultural Chemicals, Drinking Water and Public Health, An Epidemiology Review,* Soil Society of America, Workshop Paper No. 2 (1986).

[18]National Academy of Sciences, pp. 422–423.

[19]Leon F. Burmeister, et. al., "Selected Cancer Mortality and Farm Practices in Iowa, *Am. J. of Epid.* 118, No. 1 (1983), 72–77.

[20]Strange and Krupicka, pp. 41–53.

[21]Margaret M. Dorsch, et. al., "Congenital Malformations and Maternal Drinking Water Supply in Rural South Australia: A Case Control Study," *Am. J. of Epid.,* 119, No. 4 (1984), 473–485.

[22]National Academy of Sciences, p. 423.

[23]Ibid, p. 424.
[24]Council for Agricultural Science and Technology, p. 10.

11. Fluoridation
[1]Ellen Ruppel Shell, "An Endless Debate," *Atlantic,* Vol. 258, August 1986, pp. 26–31.
[2]Centers for Disease Control, *Water Fluoridation, A Manual for Engineers and Technicians* (Atlanta, GA: U.S. Public Health Service, September 1986), p. 121.
[3]C.E. Koop, U.S. Surgeon General, Letter to John W. Hernandez, Jr., Deputy Administrator, U.S. EPA, July 30, 1982.
[4]Centers for Disease Control, p. 20.
[5]John A. Yiamouyiannis, "NIDR Study Shows No Relationship Between Fluoridation and Tooth Decay Rate," *American Laboratory,* May 1989, pp. 8–10.
[6]Jayanth V. Kumar, et. al., "Trends in Dental Fluorosis and Dental Cavities Prevelences in Newburgh and Kingston, NY," *Am. J. Pub. Health,* May 1989, pp. 565–569.
[7]Mark Diesendorf, "The Mystery of Declining Tooth Decay," *Nature,* 322, 10 July 1986, 125–129.
[8]U.S. EPA, " National Primary Drinking Water Regulations; Fluoride, Final Rule," *50 Federal Register,* 14 November 1985, p. 47143.
[9]Ibid, p. 47143.
[10]Ibid, p. 47144.
[11]Ibid.
[12]Ibid.
[13]Ibid, p. 47150.
[14]John Yiamouyiannis, *Fluoride, The Aging Factor* (Delaware, OH: Health Action Press, 1986), p. 65.
[15]Dean Burk, "Fluoridation: A Burning Controversy," *Bestways,* April 1982, p. 42.
[16]Yiamouyiannis, *Fluoride, The Aging Factor,* 1986, p. 71.
[17]Dean Burk, *Increased Cancer Mortality Linked with Artificial Fluoridation of Drinking Water in Large American and British Cities: Year-By-Year Time-Trends Compared Immediately Before and After Fluoridation* (Washington, D.C.: Dean Burk Foundation, 1980).
[18]Marvin A. Schneiderman, *Fluoridation and Health: A Short Review of Some Evidence From the United States* (Bethesda, MD: National Cancer Institute, NIH, 1979).
[19]"Fluoride Might Be Tied to Cancer," *The Cleveland Plain Dealer,* 27 January 1990, p. 6C.
[20]Burk, *Bestways,* p. 42.
[21]K. Tennakone and S. Wickramanayake, "Aluminum Leaching from Cooking Utensils," *Nature,* 325, 15 January 1987, 202.
[22]Yiamouyiannis, *Fluoride, The Aging Factor,* 1986.
[23]Ibid.
[24]J. Emsley, et. al., "An Unexpectedly Strong Hydrogen bond: Ab Initio Calculations and Spectroscopic Stuides of Amide-Fluoride Systems," *J. Am. Chemical Society,* 103 (1980), 24–28; J. Emsley, D.J. Jones, and R.E. Overill, "The Uracill-Fluoride Interaction: Ab Initio Calculations Including Solvation," *J. Chem. Soc. Chem. Commun.,* 9 (1982), 476–478.
[25]Yiamouyiannis, *Fluoride, The Aging Factor,* 1986, p. 108.
[26]Ibid.
[27]W.A. Price, "Race Decline and Race Regeneration," *J. Am. Dental Assoc.,* 28 (1941), 550.
[28]P. Sadtler, "Fluorine Gases in Atmosphere as Industrial Waste Blamed for Death and Chronic Poisoning of Donora and Webster, Pennsylvania Inhabitants," *Chem. Eng. News,* 26 (1948), 3692.
[29]K. Roholm, "The Fog Disaster in the Meuse Valley 1930: A Fluorine Intoxication," *J. Ind. Hyg. Toxicol.,* 19, No. 3 (1937), 126–137.
[30]Shell, p. 29.
[31]John A. Yiamouyiannis, personal communication, November 12, 1989.

[32]Shell, p. 31.

[33]Yiamouyiannis, *Fluoride, The Aging Factor,* 1986, p. 106.

[34]Shell, p. 30.

[35]Centers for Disease Control, p. 118.

[36]U.S. EPA, "National Primary and Secondary Drinking Water Regulations; Fluoride, Final Rule," *51 Federal Register,* 2 April 1986, p. 11396–11412.

[37]Ibid, p. 11404.

12. Synthetic Organic Chemicals

[1]Testimony of Thomas C. Jorling, Assistant Administrator, U.S. EPA, before the U.S. Senate Committee on Environment and Public Works, hearings on the Safe Drinking Water Act, March 29, 1982, p. 260.

[2]U.S. International Trade Commission, *Synthetic Organic Chemicals, U.S. Production and Sales, 1988,* U.S. ITC Publication 2219 (Washington, D.C.: U.S. ITC, September 1989).

[3]Walter M. Shackelford and David M. Cline, "Organic Compounds in Water," *Environ. Sci. Technol.* 20, No. 7 (1986), 652–657.

[4]National Research Council, *Toxicity Testing* (Philadelphia: National Academy Press, 1984), p. 51.

[5]Halina S. Brown, Donna R. Bishop, and Carol S. Rowan, "The Role of Skin Absorption as a Route of Exposure for Volatile Organic Compounds (VOCs) in Drinking Water," **AJPH,** May 1984, pp. 479–483.

[6]Shackelford and Cline, 1986.

[7]National Academy of Sciences, *Drinking Water and Health,* Volume 1, Safe Drinking Water Committee, National Research Council (Washington, D.C.: National Academy of Sciences, 1977), pp. 4–6.

[8]T.A. Bellar, J.J. Lichtenberg, and R.C. Kroner, "The Occurrence of Organohalides in Chlorinated Drinking Waters," *JAWWA,* September 1974, pp. 703–706.

[9]National Cancer Institute, *Report on Carcinogenesis Bioassay of Chloroform,* Carcinogenesis Program, Division of Cancer Cause and Prevention (Bethesda, MD: NCI, March 1, 1976).

[10]U.S. EPA, "National Interim Primary Drinking Water Regulations; Control of THMs in Drinking Water," *44 Federal Register,* 29 November 1979, p. 68629.

[11]Joseph A. Cotruvo, "THMs in Drinking Water," *Environ. Sci. Technol.* 15, No. 3 (1981), 268–274.

[12]National Academy of Sciences, Vol. 1, p. 1

[13]"Regulating Organics," interview with Michael Cooke, director, U.S. EPA Office of Drinking Water, **JAWWA,** January 1987, pp. 10–23.

[14]M.J. McGuire, and R.G. Meadow, "AWWARF Trihalomethane Survey," **JAWWA,** January 1988, p. 61.

[15]U.S. EPA, *44 Federal Register,* p. 68631.

[16]National Academy of Sciences, Vol. 1.

[17]W. Emile Coleman, et. al., "GC/MS Analysis of Mutagenic Extracts of Aqueous Chlorinated Humic Acid - A Comparison of the By-Products to Drinking Water Contaminants," *Environ. Sci. Technol.* 18, No. 9 (1984), 674–681.

[18]U.S.EPA, *44 Federal Register,* p. 68629.

[19]K. Cantor, et. al., "Association of Cancer Mortality with Trihalomethanes in Drinking Water," **JNCI,** 61 (1977), 979; M.D. Morgan, et. al., "Drinking Water Supplies and Various Specific Cancer Mortality Rates, *Journal of Env. Pathology and Toxicology,* 2 (1979), 873; S.R. Zieler, et. al., "Type of Disinfectant in Drinking Water and Patterns of Mortality in Massachusetts," *Environ. Health Persp.,* 69, (1986), 275–279.

[20]H.J. Kool, C.F. van Kreiji, and B.C.J. Zoetman, "Toxicology Assessment of Organic Compounds in Drinking Water," **CRC Critical Reviews in Environmental Control,** 12, No. 4, 307–357.

[21]Marise Gotlieb and Jean Carr, "Case-Control Cancer Mortality Study and Chlorination of Drinking Water in Louisianna," *Environ. Health Persp., 46* (1982), 169–177; R.H. Harris, *The Implications of Cancer-Causing Substances in Mississippi River Water* (Washington, D.C.: Environmental Defense Fund, 1974).

[22]National Academy of Sciences, *Drinking Water and Health Disinfectants and Disinfectant Byproducts,* Volume 7, Safe Drinking Water Committee, National Research Council (Washington, D.C.: National Academy Press, 1987), p. 58; Kenneth P. Cantor, et. al., "Bladder Cancer, Drinking Water Source, and Tap Water Consumption: A Case-Control Study," *JNCI,* 79, No. 6, December 1987, 1269–1279.

[23]Ronald Kuzma, Cecilia Kuzma, and C. Ralph Buncher, "Ohio Drinking Water Source and Cancer Rates," *Am. J. Pub. Health,* 67 (1977), 725–729.

[24]S. Maruoka and S. Yamanaka, "Production of Mutagenic Substances by Chlorination of Water," *Mutagen Research,* 79 (1980), 381–386; J. Rook, "Formation of Haloforms During Chlorination of Natural Waters," *Water Treatment Exam.,* 23 (1974), 234–243; B. Glatz, et. al., "Examination of Drinking Water for Mutagenic Activity," *JAWWA,* 70 (1978), 465–468.

[25]John R. Meier, et. al., "Identification of Mutagenic Compounds Formed During Chlorination of Humic Acid," *Mutation Research,* 157 (1985), 111–122.

[26]National Academy of Sciences, Vol. 7, p. 197.

[27]U.S. EPA, *44 Federal Register,* p. 68625.

[28]National Academy of Sciences, Vol. 1, p. 492.

[29]James J. Westrick, Wayne Mello, and Robert F. Thomas, "The Groundwater Supply Survey," *JAWWA,* May 1984, pp. 52–59.

[30]U.S. EPA, "National Primary Drinking Water Regulations — Synthetic Organic Chemicals; Monitoring for Unregulated Contaminants; Final Rule," *Federal Register,* 8 July 1987, p. 25702.

[31]National Academy of Sciences, Vol. 1, p. 492.

[32]Mark Jones, et. al., *Chemistry, Man and Society,* 4th ed. (Philadelphia: Saunders College Publishing, 1983), p. 358.

[33]R.M. Harrison, R. Perry, and R.A. Wellings, "Review Paper: Polynuclear Aromatic Hydrocarbons in Raw, Potable, and Waste Waters," *Water Research,* 9 (1975), 331–346.

[34]Nancy K. Kim, and Daniel W. Stone, *Organic Chemicals in Drinking Water* (Albany: New York State Dept. of Health, April 1981), p. 11.

[35]Ronald Hites, et. al., "Potentially Toxic Organic Compounds in Industrial Wastewaters and River Systems: Two Case Studies," In: *Monitoring Toxic Substances,* American Chemical Society, 1979, pp. 63–90.

[36]Per Larsson, Anders Thurén, and Gunnar Gahnström, "Phthalate Esters Inhibit Microbial Activity in Aquatic Sediments," *Environ. Pollution,* Series A, 4 (1986), 223–231.

[37]U.S. EPA, *Ambient Water Quality Criteria for Phthalate Esters,* EPA 44015-80-067 (Washington, D.C.: Criteria and Standards Division, October 1980), p. A-3.

[38]C.S. Giam, et. al., "Phthalate Ester Plasticizers: A New Class of Marine Pollutants," *Science,* 199, No. 27 (1978), 419–421.

[39]K.F. Sullivan, E.L. Atlas, and C.S. Giam, "Adsorption of Phthalate Acid Esters from Sea Water," *Environ. Sci. Technol.,* 16 (1982), 428–432.

[40]J.C. Peterson, and D.H. Freeman, "Phthalate Ester Concentration Variations in Dated Sediment Cores form Chesapeake Bay," *Environ. Sci. Technol.,* 16 (1982), 464–469.

[41]Ibid.

[42]U.S. EPA, *Ambient Water Quality Criteria for Phthalate Esters,* p. C-3.

[43]Kim and Stone, p. 11.

[44]P.R. Graham, "Phthalate Ester Plasticizers - Why and How They are Used," *Environ. Health Persp.,* January 1973, pp. 3–12; *U.S. Tariff Commission Report on Plasticizers* (Washington, D.C.: U.S. Government Printing Office, 1971).

[45]Walter J. Kozumbo, Rosanna Kroll and Robert J. Rubin, "Assessment of the Mutagenicity of Phthalate Esters," *Environ. Health Persp.,* 45 (1982), 103–109.

[46]Prahlad K. Seth, "Hepatic Effects of Phthalate Esters," *Environ. Health Persp.*, 45 (1982), 27–34.

[47]W.M. Kluwe, et. al., "Carcinogenicity Testing of Phthalate Esters and Related Compounds by the National Toxicology Program and the National Cancer Institute," *Environ. Health Persp.*, 45 (1982), 129–133.

[48]U.S. EPA, *Drinking Water Quality Criteria for Chlorinated Phenols,* EPA 440/5-80-032 (Washington, D.C.: Criteria and Standard Division, October 1980), p. A-1.

[49]William D. Rowe, "The Nature and Pollution Sources of Phenol," In: *Evaluation Methods for Environmental Standards,* Fredirick Hagman, ed. (Boca Raton, FL: CRC Press Inc., 1983), pp. 59–60.

[50]U.S EPA, *Drinking Water Criteria for Chlorinated Phenols,* p. A-7.

[51]Ibid, p. A-8.

[52]Rowe, p. 59–60

[53]K.C. Swallow, N.S. Shifrin, and P.J. Doherty, "Hazardous Organic Compound Analysis," *Environ. Sci. Technol.*, 22, No. 2 (1988), 136–142.

13. Pesticides

[1]Council on Environmental Quality, *Integrated Pest Managment* (Washington, D.C.: U.S. Government Printing Office, 1979).

[2]U.S. EPA, *Pesticide Industry Sales and Usage 1985 Market Estimates,* Economic Analysis Branch (Washington, D.C.: Office of Pesticide Programs, September 1986).

[3]Ibid.

[4]Council on Environmental Quality, p. 3.

[5]Economic Research Service, *Agricultural Resources Situation and Outlook,* (Washington, D.C.: U.S. Dept. of Agriculture, January 1988).

[6]Council on Environmental Quality, p. 1.

[7]Ibid.

[8]Kenneth Cantor and Aaron Blair, "Agricultural Chemicals, Drinking Water, and Public Health: An Epidemiologic Overview," Sail Science Society of America, Workshop Paper No. 2, 1986.

[9]Staff Report, *EPA Pesticide Regulatory Program Study,* Subcommittee on Department Operations, Research and Foreign Agriculture, U.S. House of Representatives, 98th Congress, 2nd Session, December 17, 1982.

[10]National Research Council, *Toxicity, Strategies to Determine Needs and Priorities* (Washington, D.C.: National Academy Press, 1984).

[11]U.S. General Accounting Office, *Pesticides: EPA's Formidable Task to Assess and Regulate Their Risks,* GAO/RCED-86-125 (Washington, D.C.: U.S. GAO, April 1986), p. 25.

[12]U.S. General Accounting Office, *Federal Pesticide Registration Program: Is It Protecting the Public and the Environment Adequately From Pesticide Hazards?,* RED-76 (Washington, D.C.: U.S. GAO, December 4, 1975).

[13]U.S. EPA, *Regulatory Impact Analysis: Data Requirements for Registering Pesticides Under the Federal Insecticide, Fungicide and Rodenticide Act* (Washington, D.C.: Office of Pesticide Programs, 1982), p. 27.

[14]The Conservation Foundation, *Groundwater Protection* (Washington, D.C.: The Conservation Foundation, 1987), p. 147.

[15]Maureen K. Hinkle, "Problems with Conservation Tillage," *J. Soil & Water Conservation,* May–June 1983, pp. 201–206.

[16]Mary Barnett, "A Better Way to Fight Bugs," Amicus Journal, Spring 1983, pp. 27–31.

[17]Sandra Postel, "Defusing the Toxics Threat: Controlling Pesticides and Industrial Waste," *Worldwatch Paper 79,* September 1987.

[18]P.L. Adkisson et. al., "Controlling Cotton's Insect Pests: A New System," *Science,* 216 (1982), 19–22.

[19]Marty Strange, Liz Krupicka, and Dan Looker, "The Hidden Health Effects of Pesticides," In: *It's Not All Sunshine and Fresh Air* (Walthill, Nebraska: Center for Rural Affairs, April 1984), pp. 55–75.

[20]Samuel S. Epstein, "National Pesticide Policy Is Dangerous to Your Health," *The New York Times,* 1 December, 1984.

[21]Council on Environmental Quality, pp. 67–86.

[22]"Reducing Pesticide Use," *The Baltimore Sun,* March 13, 1985.

[23]Samuel Epstein, "National Pesticide Policy Is Dangerous to Your Health".

[24]David Anderson and Ronald Hites, "Chlorinated Pesticides in Indoor Air," *Environ. Sci. Technol.,* 22, No. 6, 717–720.

[25]Ruth A. Lowengart, et. al, "Childhood Leukemia and Parent's Occupational and Home Exposures," *JNCI,* 79, No. 1 (1987), 39–46.

[26]"Pesticides Blamed in Birth Defects," *Am. Med. News,* February 22, 1980.

[27]Laura Weiss, *Keep Off the Grass* (Washington, D.C.: Public Citizen's Congress Watch, April 1989).

[28]Chris Chinlund, "Green But Toxic, Use of Lawn Pesticides Concerns State Officials," *The Boston Globe,* 28, June 1984, p. 15.

[29]Weiss, p. 5.

[30]U.S. General Accounting Office, *Nonagricultural Pesticides: Risks and Rgulation,* GAO/RCED-86-97 (Washington, D.C.: U.S. GAO, April 1986), p. 17.

[31]Statement of Peter G. Guerro, Associate Director Environmental Protection Issues, Resources, Community, and Economic Development Division, before the Subcommittee on Toxic Substances, Environmental Oversight, Research and Development of the U.S. Senate Committee on Environment and Public Works, in: U.S. GAO, *Reregistration and Tolerance Reassessment Remain Incomplete for Most Pesticides,* GAO/T-RCED-89-40 (Washington, D.C.: U.S. GAO, May 15, 1989).

[32]U.S. General Accounting Office, *Nonagricultural Pesticides, Risks and Regulation,* GAO/RCED86-87 (Washington, D.C.: U.S. GAO, April 1986).

[33]Ibid, p. 21.

[34]U.S. GAO, *Nonagricultural Pesticides, Risks and Regulation,* p. 31.

[35]Ibid.

[36]U.S. General Accounting Office, *Pesticides: EPA's Formidable Task to Assess and Regulate Their Risks,* GAO/RCED -86-125 (Washington, D.C.: U.S. GAO, April 1986), p.85.

[37]Testimony of Laurie Mott, Natural Resources Defense Council, before the Subcommittee on Health and the Environment, U.S. House Committee on Energy and Commerce, June 8, 1987.

[38]Testimony of Albert Meyerhoff and Nancy Drabble, Campaign for Pesticide Reform, before the Subcommittee on Department Operations, Research and Foreign Agriculture, U.S. House Committee on Agriculture, hearings on Reauthorization of FIFRA, March 19, 1986.

[39]U.S. EPA, *Pesticides in Groundwater Data Base 1988 Interim Report* (Washington, D.C.: Office of Pesticide Programs, U.S. EPA, December 1988).

[40]U.S. EPA, *Pesticides in Groundwater Background Document* (Washington, D.C.: U.S. EPA, May 1986), p. 9.

[41]E.G. Nielsen and L.K. Lee, *Potential Groundwater Contamination from Agricultural Chemical: A National Perspective* (Washington, D.C.: Economic Research Service, U.S. Dept. of Agriculture, 1987), p. 9.

[42]Statement of Gary Englund, Chief, Section of Water Supplies, Minnesota Department of Health, before the Subcommittee on Toxic Substances and Environmental Oversight, U.S. Senate Committee on Environmental Public Works, hearings on Pesticides and Groundwater, September 9, 1986, p. 52.

[43]Statement of Richard D. Kelley, Environmental Specialist, Iowa Department of Water, Air and Waste Management, before the Subcommittee on Department Operations, Research and Foreign Agriculture, U.S. House Committee on Agriculture, hearings on FIFRA; and Pesticide Import and Export Act of 1985,May 21, 1985.

[44]Nielsen and Lee, p.25.

[45]"Pesticides: Balancing Risks and Benefits," **Health and Environment Digest,** Vol. 1, No. 1 (1987).

[46]U.S. EPA, **Pesticides in Groundwater Background Document,** May 1986, p. 45.

[47]Shelia Hoar, et. al., "Agricultural Use and Risk of Lymphona and Soft-Tissue Sarcoma," JAMMA, 256, No. 9 (1986), 1141–1147.

[48]A. Blair, and T.L. Thomas, "Leukemia Among Nebraska Farmers: A Death Certificate Study," **Am. J. of Epidem.,** 110 (1979), pp. 26–273.

[49]Statement of Leon F. Burmeister, Professor, Department of Preventive Medicine and Environmental Health, University of Iowa, before the Subcommittee on Department Operations, Research and Foreign Agriculture, U.S. House Committee on Agriculture, hearings on FIFRA; and Pesticide Import and Export Act of 1985, May 20, 1985.

[50]Ibid, p. 389.

[51]A. Blair, J.F. Fraumeeni, Jr., and T.J. Mason, "Geographic Patterns of Leukemia in the U.S.," **J. Chronic Dis.,** 33 (1980), 521–526.

[52]Blair and Thomas, 1979; L.F. Burmeister, S.F. Van Lier, and P. Isacson, "Leukemia and Farm Practices in Iowa," **Am. J. of Epidem.,** 115 (1982), 720–728.

[53]Strange, Krupicka, and Looker, p. 56.

[54]U.S. EPA, **Pesticides in Groundwater Background Document,** p. 6.

[55]T.R. Eichers, "Farm Pesticide Economic Evaluation,"U.S. Department of Agriculture Economics and Statistics Service, Agriculture Economic Report, No. 464, 1981.

[56]U.S. Dept. of Agriculture, **1980 Appraisal Part I: Soil, Water and Related Resources in The United States: Status, Condition, and Trends** (Washington, D.C.: U.S. Dept. of Agriculture).

[57]Nielsen and Lee, p. 3.

[58]Stanley E. Manahan, **Environmental Chemistry,** 3rd ed. (Boston: Willard Grant Press, 1979), p. 175.

[59]Suzanne Hively, "Pesticides Cut Back in Plant-Care Plan," **The Plain Dealer,** 13 March 1988, p. 11G.

[60]Sandra Postel, "Defusing the Toxics Threat: Controlling Pesticides and Industrial Waste," **Worldwatch Paper 79,** The Worldwatch Institute, September 1987.

[61]Devorah Lanner and Chuck Hassebrook, **Low Input Sustainable Family Farm Agriculture,** Prepared for the Midwest Sustainable Agriculture Working Group (Washington, D.C.: Center for Rural Affairs, March, 1989).

[62]Charles C. Geisler, et. al, "Sustained Land Productivity: Equity Consequences of Alternative Agricultural Technologies," In: **The Social Consequences and Challenges of New Agricultural Technologies,** Gigi M. Bernarddi and Charles Geisler, eds. (Boulder: Westview Press, 1984), cited in Lanner and Hassebrook, p. 2.

14. Sources of Chemical Contaminants

[1]Statement of Honorable Dave Durenberger, before the Subcommittee on Toxic Substances and Environmental Oversight, U.S. Senate Committee on Environment and Public Works, hearings on Groundwater Contamination and Protection, June 17, 1985, p. 1.

[2]Jane L. Bloom, "The RCRA Amendments and Groundwater," **Environment,** Jan./Feb. 1985, pp. 44–45.

[3]U.S. EPA, **Report to Congress on Injection of Hazardous Waste,** EPA 570/9-85-003 (Washington, D.C.: Office of Drinking Water, May 1985), p. 5.

[4]Ibid, p. 6.

[5]U.S. Office of Technology Assessment, **Protecting the Nation's Groundwater From Contamination, Volume II,** OTA-0-276 (Washington, D.C.: U.S. OTA, October 1984).

[6]Ibid.

[7]R. Patrick, E. Ford, and J. Quarles, **Groundwater Contamination in the United States,** 2nd Ed. (Philadelphia: University of Pennsylvania Press, 1987), p. 72.

[8]R. Patrick, E. Ford, and J. Quarles, *Groundwater Contamination in the United States,* 2nd Ed. (Philadelphia: University of Pennsylvania Press, 1987), p. 75.

[9]D.E. Burmaster, and R.H. Harris, "Groundwater Contamination: An Emerging Threat," *Technology Review,* 85, No. 5 (1982), 50–62.

[10]U.S. EPA, *Report to Congress: Waste Disposal Practices and Their Effects on Groundwater,* EPA 570/9-77-001 (Washington, D.C.: U.S. EPA Office of Water Supply, June 1977).

[11]David Adams, "Fears Rise Over Deep-Well Wastes," *Akron Beacon Journal,* 19 November 1989, p. 1.

[12]U.S. EPA, *Protection of Public Water Supplies from Groundwater Contamination,* Seminar Publication, EPA 625/4-85/016 (Cincinnati, OH: U.S. EPA, September, 1985), p. 116.

[13]Bloom, p. 45.

[14]Water Resources Program, Princeton University, *Groundwater Contamination From Hazardous Wastes* (Englewood Cliffs, NJ: Prentice-Hall, Inc., 1984), p. 144.

[15]U.S. OTA, *Protecting the Nation's Groundwater from Contamination,* p. 271.

[16]Ibid.

[17]U.S. EPA, *Surface Impoundment Assessment National Report,* EPA 570/9-84-002 (Washington, D.C.: Office of Drinking Water, December 1983).

[18]Ibid, p. 5.

[19]U.S. EPA, *Report to Congress: Waste Disposal Practices and Their Effects on Groundwater,* p. 314.

[20]J.A. Cotteral and D.P. Norris, "Septic Tank Systems," *ASCE Journal,* Sanitary Engineering Division, 95, No. SA4 (1969), 715–746.

[21]A. Hershaft, "The Plight and Promise of On-Site Wastewater Treatment," *Compost Science,* 17, No. 5 (1976), 6–13.

[22]Larry Canter and Robert C. Knox, *Evaluation of Septic Tank System Effects on Groundwater Quality,* Project Summary, EPA 600/52-84-109 (Washington, D.C.: U.S. EPA, September 1984).

[23]U.S. EPA, *Waste Disposal Practices and Their Effects on Groundwater,* p. 186.

[24]Canter and Knox, *Evaluation of Septic Tank System Effects on Groundwater Quality.*

[25]*Larry Canter and Robert C. Knox, Septic Tank System Effects on Groundwater Quality* (Chelsea, MI: Lewis Publishers Inc., 1985), p. 2.

[26]Ibid, p. 2.

[27]Canter and Knox, *Septic Tank System Effects on Groundwater Quality,* p. 13.

[28]U.S. EPA, *Waste Disposal Practices and Their Effects on Groundwater,* p. 295.

[29]Henry Hughs, James Pike, and Keith Porter, *Assessment of Groundwater Contamination by Nitrogen and Synthetic Organics in Two Water Districts in Nassau County, New York,* Water Resources Program, Center for Environmental Research (Ithaca, NY: Cornell University, January 1985), pp. 1–52.

[30]Foppe B. DeWalle, "Determination of Toxic Chemicals in Effluent from Household Septic Tanks," Project Summary, EPA/600/52-85/050 (Cincinnati, OH: U.S. EPA, August 1985).

[31]Canter and Knox, *Septic Tank System Effects on Groundwater Quality,* p. 82; O.B. Kaplan, "Some Additives to Septic Tank Systems May Poison Groundwater," *J. Environ. Health,* March/April 1983, p. 259.

[32]Burmaster and Harris, p. 58.

[33]U.S. Office of Technology Assessment, *Technologies and Management Strategies for Hazardous Waste Control,* OTA-M-196 (Washington, D.C.: U.S. GPO, June 1983).

[34]U.S. OTA, *Protecting the Nation's Groundwater from Contamination,* p. 287.

[35]Peggy Rodgers, U.S. EPA, personal communication, February 1988.

[36]U.S. OTA, *Protecting the Nation's Groundwater from Groundwater,* p. 278.

[37]Versar, Inc., *Summary of State Reports on Releases from Underground Storage Tanks,* EPA/600/M-86/020 (Washington, D.C.: Office of Underground Storage Tanks, U.S. EPA, August 1986), pp. 1-4–1-5.

[38]*Here a Tank, There a Tank, Everywhere a . . . , Tank Removal and Abandonment,* L.U.S.T. Line, New England Interstate Water Pollution Control Commission, Bulletin No. 7, December 1987, p. 1.

[39]U.S. OTA, *Protecting the Nation's Groundwater from Groundwater,* p. 279.

[40]Statement of Paul M. Yaniga, before the Subcommittee on Toxic Substances and Environmental Oversight, U.S. Senate Committee on Environment and Public Works, hearings on Hydrocarbon Contamination of Groundwater: Assessment and Abatement, March 1, 1984.

[41]Marjorie Sun, "Groundwater Ills: Many Diagnoses, Few Remedies," *Science,* 232 (1986), 1490–1493.

[42]Joel Millman, "Tank Trouble," *Technology Review,* Feb/March 1986, pp. 74–75.

[43]U.S. OTA, *Protecting the Nation's Groundwater from Contamination,* p. 280.

[44]David Scheel, "Road Salt Contaminates Well, Causes Health Hazard," *J. of Environ. Health,* 47, No. 4, 202–203.

[45]The Conservation Foundation, *Groundwater Protection* (Washington, D.C.: The Conservation Foundation, 1987), pp. 162–163.

[46]H. Bowwer, *Groundwater Hydrology* (New York: McGraw Hill, Inc., 1978).

[47]Patrick, et. al., pp. 77–78.

[48]U.S. EPA, *Protection of Public Water Supplies from Groundwater Contamination,* p. 113.

[49]Ibid.

[50]U.S. Department of Transportation, Research and Special Programs Administration, *Annual Report on Pipeline Safety, Calendar Year 1988* (Washington, D.C.: U.S. DOT, 1988), p. iii.

[51]Ibid, pp. 51–55.

[52]U.S. OTA, *Protecting the Nation's Groundwater from Contamination,* p. 282.

[53]Elizabeth Sullivan, "Thousands Flee Toxic Spills Path," *The Plain Dealer,* 19 February 1988, p. 1.

[54]U.S. OTA, *Protecting the Nation's Groundwater from Contamination,* p. 282.

[55]U.S. General Accounting Office, *Illegal Disposal of Hazardous Waste: Difficult to Detect or Deter,* GAO/RCED-85-2 (Washington, D.C.: U.S. GAO, February 22, 1985).

[56]Savant Associates Inc., Response Analysis Corp for EPA, *Experience of Hazardous Waste Generators With EPA's Phase I RCRA C Program* (Washington, D.C.: U.S. EPA, September 1, 1983).

[57]U.S. GAO, *Illegal Disposal of Hazardous Waste: Difficult to Detect or Deter,* p. 22.

[58]Ibid, p. 23.

[59]Ibid, p. 37.

15. Inexpensive Screening Tests

[1]Ronald C. Dressman, and Alan A. Stevens, "The Analysis of Organohalides in Water — An Evaluation Update," *JAWWA,* August 1983, pp. 431–434.

[2]Am. Pub. Health Assoc., Am. Water Works Assoc., and Water Pollution Control Federation, *Standard Methods for the Examination of Water and Wastewater,* 16th ed. (Washington, D.C.: APHA, 1985), p. 516.

[3]Alan A. Stevens, et. al., "Organic Halogen Measurements: Current Uses and Future Prospects," *JAWWA,* April 1985, pp. 146–154.

[4]Am. Pub. Health Assoc., et. al., p. 518.

[5]Ibid, p. 518.

[6]Stevens, et. al, p. 151.

[7]Am. Pub. Health Assoc., et. al., p. 507.

[8]Michael J. Barcelona, *TOC Determinations in Groundwater,* Water Supply Division, Illinois Dept. of Energy and Natural Resources, 1983.

16. Testing Your Water

[1]Cliff J. Kirchmer, "Quality Control in Water Analysis," *Environ. Sci. Technol.*, 17, No. 4 (1983), 174A–181A.

[2]U.S. EPA, "Guidelines Establishing Test Procedures for the Analysis of Pollutants Under the Clean Water Act; Final Rule and Interim Final Rule," **49 Federal Register,** 26 October 1984, p. 43237.

[3]Arnold E. Greenberg and William J. Hausler, Jr., "Water Laboratory Certification," *Environ. Sci. Technol.*, 12, No. 5 (1981), 520–522.

[4]Kirchmer, p. 181A.

[5]Am. Pub. Health Assoc., Am. Water Works Assoc., and Water Pollution Control Fed., **Standard Methods for the Examination of Water and Wastewater,** 16th ed. (Washington, D.C.: APHA, 1985), p. 518.

[6]Testimony of Pat Bourez before the U.S. House Committee on Government Operations, June 22, 1983, p. 109.

[7]National Cancer Institute, **Human Health Considerations of Carcinogenic Organic Chemical Contaminants in Drinking Water,** National Institute of Health Position Paper (Bethesda, MD: U.S. Dept. of Health, Education and Welfare, April 1978).

[8]U.S. EPA, "National Interim Primary Drinking Water Regulations; Control of Trihalomethanes in Drinking Water," **44 Federal Register,** 29 November 1979, p. 68669.

[9]Irving J. Selikoff, "Living in a Chemical World," **Bull. Environ. Contam. Toxicol.** (1984), pp. 33682–33695.

17. Home Treatment Methods and Devices

[1]Benjamin W. Lykins, et. al., **Granular Activated Carbon for Removing Non-trihalomethane Organics from Drinking Water,** Project Summary, EPA-600/52-84-165 (Cincinnati, OH: U.S. EPA, December 1984).

[2]Frank A. Bell, et. al., "Studies on Home Water Treatment Systems," **JAWWA,** April 1984, pp. 126–130.

[3]U.S. EPA, **Protection of Public Water Supplies from Groundwater Contamination,** Seminar Publication, EPA/625/4-85/016 (Washington, D.C.: U.S. EPA, September 1985), p. 128.

[4]Ibid.

[5]Lykins, et. al., p. 2.

[6]S. Monarca, J.R. Meier, and R.J. Bull, "Removal of Mutagens from Drinking Water by Granular Activated Carbon," **Water Research,** 17, No. 9 (1983), 1015–1026.

[7]Jerry D. Lowry and Sylvia B. Lowry, "Modeling POE Radon Removal by GAC," **JAWWA,** October 1987, pp. 85–88.

[8]Frank Bell, Ervin Bellack, and Joseph Cotruvo, "Water Quality Improvement in the Home," unpublished paper, April 1984.

[9]P. Regunathan, W.H. Beauman, and E.G. Kreusch, "Efficiency of Point-of-Use Treatment Devices," **JAWWA,** January 1983, pp. 42–50.

[10]U.S. EPA, **Third Phase/Update: Home Drinking Water Treatment Units Contract,** Criteria and Standards Division (Washington, D.C.: Office of Drinking Water, March 1982), p. 2.

[11]Lee Rozelle, "Point-of-Use and Point-of-Entry Drinking Water Treatment," **JAWWA,** October 1987, p. 53–59.

[12]"Carbon Filters: Bigger is Better," **Consumer Reports,** January 1990, pp. 33–35.

[13]Craig Wallis, Charles H. Stagg and Joseph L. Melnick, "The Hazards of Incorporating Charcoal Filters Into Domestic Water Systems," **Water Research,** 8 (1974), 111–113.

[14]R.S. Tobin, D.K. Smith, and J.A. Lindsay, "Effects of Activated Carbon and Bacteriostatic Filters on Microbiological Quality of Drinking Water," **Appl. Envir. Micro.,** March 1981, p. 6; Edwin E. Geldrich, et. al., "Bacterial Colonization of Point-of-Use Water Treatment Devices," **JAWWA,** February 1985, pp. 72–80.

[15]U.S. EPA, **Third Phase/Update,** p. 2.

[16]Am. Pub. Health Assoc., Am. Water Works Assoc., and Water Pol. Control Fed., *Standard Methods for the Examination of Water and Wastewater,* 16th ed. (Washington D.C.: APHA, 1985), p. 857.

[17]Geldrich, et. al., 1985, p. 77; Raymond H. Taylor, et. al., "Testing of Home Use Carbon Filters," **JAWWA,** October 1979, p. 577.

[18]Tobin, et. al., p. 650.

[19]Randy A. Dougherty, **NSF Standards and the Listing Program for Drinking Water Treatment Units,** RAD-6/5/86 (June 1986), pp. 1–5.

[20]Geldreich, et. al., 1985, p. 79.

[21]Bell, et. al., "Water Quality Improvement in the Home," p. 7.

[22]Ibid.

[23]Donald Reasoner, et. al., "Microbiological Characteristics of Third-Faucet Point-of-Use Devices," **JAWWA,** October 1987, pp. 60–66.

[24]Ibid, p. 66.

[25]Theodore B. Shelton and Milton Bogen, **Home Drinking Water Treatment Technologies and Devices: Mechanical Filtration/Carbon Filtratiion/Revese Osmosis/Distillation,** No. 47, Cooperative Extension Service (New Brunswick, NJ: Cook College, Rutgers State University, 1985), p. 4.

[26]Edward S.K. Chian, Willis N. Bruce, and Herbert H.P. Fang, "Removal of Pesticides by Reverse Osmosis," **Environ. Sci. Technol.,** 9, No. 1 (1975), 52–59.

[27]Regunthan, p. 44.

[28]Ibid, p. 43.

[29]Bell, et. al., **JAWWA,** 1984, p. 130.

[30]"Reverse-Osmosis Systems: Slow But Effective," **Consumer Reports,** January 1990, pp. 36–38.

[31]Shelton and Bogen, p. 4.

[32]Rozelle, pp. 55–56.

[33]"Reverse-Osmosis: Slow But Effective," p. 38.

[34]Rozelle, pp. 55–56.

[35]Shelton and Bogen, p. 5.

[36]Eldon Muehling, Pure Water, Inc., personal communication, August 1989.

[37]"Distillers: A Drop In the Bucket," **Consumer Reports,** January 1990, pp. 39–40.

[38]R.S. Tobin, et. al., "Methods for Testing the Efficiency of Ultraviolet Light Disinfection Devices for Drinking Water," **JAWWA,** September 1983, p. 484.

[39]Ibid, pp. 481–484.

[40]Envir. Health Director, **Laboratory Testing of Point-of-Use Ultraviolet Drinking Water Purifies,** Publ. 79-EHD-33 (Ottawa, Canada: Dept. Nat'l Health & Welfare, April 1979).

[41]Richard Tobin, "Testing and Evaluating Point-of-Use Treatment Devices in Canada," **JAWWA,** October 1987, pp. 43–45.

[42]William L. Lappenbusch, **Contaminated Drinking Water and Your Health** (Alexandria, VA: Lappenbusch Environmental Health Inc., 1986), p. 53.

[43]U.S. EPA, "Drinking Water Regulations; Maximum Contaminant Level Goals and National Primary Drinking Water Regulations for Lead and Copper; Proposed Rule, **Federal Register,** 18 August 1988, p. 31533.

[44]"Creating a Market: The Selling of Water Safety," **Consumer Reports,** January 1990, pp. 27–29.

[45]R. Kent Sorrell, "In-Home Treatment Methods for Removing Volatile Organic Chemicals," **JAWWA,** May 1985, pp. 72–78.

[46]Ibid, p. 75.

[47]Ibid, p. 77.

18. Bottled Water

[1]John Rossant, "Perrier's Unquenchable U.S. Thirst," **Business Week,** 29 June 1987, p. 46.

[2]Joe Studlick, and Richard Bain, "Bottled Water: Expensive Ground Water," *Water Well Journal,* July 1980, pp. 75–79.

[3]Susan Milius, "Tapped Out," *Organic Gardening,* April 1988, pp. 88–95.

[4]"What's What in Bottled Water," *Consumer Reports,* September 1980, p. 533; "How to Classify the Species of H$_2$O," *Consumer Reports,* January 1987, p. 43.

[5]Studlick and Bain, p. 76.

[6]*Consumer Reports,* 1987, p. 45.

[7]Kenneth Markussen, and Richard Bonczek, *Organic Chemical Compounds in Bottled Water Products Distributed in New York,* New York Department of Health, Bureau of Public Water Supply Protection, Operations Section, October 1982.

[8]*Consumer Reports,* 1987, p. 44.

[9]R. Rice, and B. Miller, "Structure and Regulation of the European Bottled Water Industry," In: *Safe Drinking Water* (Ann Arbor, MI: Lewis Publishers, 1985).

19. The Legal Game

[1]"Woburn Case May Spark Explosion of Lawsuits," *Science,* 234, 4 October 1986, 418–420.

[2]Frank P. Grad, "Compensating Toxic-Waste Victims," *Technology Review,* October 1985, pp. 48–50.

[3]U.S. General Accounting Office, *Report to Congress, Hazardous Waste: Issues Surrounding Insurance Availability* (Washington, D.C.: U.S. GAO, October 1987), p.42.

[4]Ibid, p. 49.

[5]Ibid, p. 44.

[6]Ibid, p. 46.

[7]Vince Sikora, "Law for Environmentalists," *J. Environ. Health,* January/February, 1985, p. 213.

[8]Ibid; U.S. EPA, "Notice Requirements for Citizen Suits Under the Safe Drinking Water Act; Final Rule," *54 Federal Register,* 12 May 1989, p. 20770; U.S. EPA, "Requirements for Citizen Suits Under the Clean Water Act; Proposed Rule," *54 Federal Register,* 30 August 1989, p. 36020.

20. Minimizing Waste and Pollution

[1]Barbara Hogan, "All Baled Up and No Place to Go," *The Conservationist,* January/February 1988, pp. 36–39.

[2]National Solid Wastes Management Assoc., *Public Attitudes Toward Garbage Disposal* (Washington, D.C.: NSWMA, May 1989).

[3]Cynthia Pollock, "Mining Urban Wastes: The Potential for Recycling," *Worldwatch Paper 76,* Worldwatch Institute, April 1987, p. 15.

[4]Joseph F. Sullivan, "States Are Making Recycling a Must," *New York Times,* January 1987, sec. 4, p. 32.

[5]Michael A. Fuoco, "Demand Grows for Shrinking Landfills," *Pittsburgh Post Gazette,* 19 October 1987, p. 1.

[6]Mary Grace Poidomani, "Landfills Here Are A Bargain," *Akron Beacon Journal,* 13 December 1987, p. 1.

[7]National Solid Wastes Management Assoc.

[8]Betsy Lammerding, "Waste Not, Want Not, More Recycling Needed to Avoid Burying Ourselves in Trash," *Akron Beacon Journal,* 5 November 1989, p. J8.

[9]Tim Darnett, "The Politics of Dumping Garbage," *American City and County,* May 1988, pp. 41–44.

[10]Garrett Hardin, "The Tragedy of the Commons," *Science,* 13 December 1968, pp. 1243–1248.

[11]Sullivan, p. 32.

[12]John Schwartz, et. al., "Turning Trash Into Hard Cash," *Newsweek,* 14 March 1988, pp. 36–37.

[13]Pollock, p. 14.

[14]G. Tyler Miller, *Living in the Environment,* 5th ed (Belmont, CA: Wadsworth Publishing Co., 1988), p. 500.

[15]Pollock, p. 22.

[16]Robert Cowles Letcher and Mary T. Sheil, "Source Separation and Citizen Recycling," In: *The Solid Waste Handbook,* William D. Robinson, ed. (New York: John Wiley and Sons, 1986), pp. 215–258.

[17]Lester R. Brown and Pamela Shaw, "Six Steps to a Sustainable Society," *Worldwatch Paper 48,* Worldwatch Institute, March 1982, pp. 35–36.

[18]A.H. Purcell, "The World's Trashiest People," *The Futurist,* Feb. 1981, pp. 51–59.

[19]Letcher and Sheil, p. 220.

[20]Ibid, p. 220.

[21]Dave Tomten, U.S. EPA Office of Solid Waste, Washington, D.C., personal communication, July 14, 1989.

[22]"Recycling, Reuse, Repairs," *Science,* 202 (1978) 34–35.

[23]U.S. EPA, "Guideline for Federal Procurement of Paper and Paper Products Containing Recovered Materials, Proposed Rule," *50 Federal Register,* 9 April 1985, p. 14076.

[24]Miller, p. 500.

[25]U.S. EPA, *50 Federal Register,* p. 14076.

[26]Pollock, p. 22.

[27]Citizen's Clearinghouse for Hazardous Waste, *Action Bulletin #21,* February 1989, p. 8.

[28]Pollock, p. 23.

[29]David Wann, "Timber and Tourists: Idaho Confronts Logging Issues," *EPA Journal,* December 1987, pp. 20–22.

[30]Pollock, p. 10.

[31]Citizen's Clearinghouse for Hazardous Waste, p. 8.

[32]Joan Nelson-Horchler, "Recycling, Plastics New Weapon," *Industry Week,* 4 July 1988, p. 54.

[33]Ibid.

[34]Ibid, pp. 54–56.

[35]Gary Chamberlain, "Recycled Plastics, Building Blocks of Tomorrow," *Design News,* 4 May 1987, pp. 51–58.

[36]"Firm Has New Uses for Plastic Trash," *The Cleveland Plain Dealer,* 1 January 1989, p. 7e.

[37]Pat Wittig, "Persistent Peril," *Organic Gardening,* February 1989, pp. 66–72.

[38]Chamberlain, p., 58.

[39]Horchler, p. 54.

[40]Alyssa A. Lappen, "Plastic Warnings," *Forbes,* 10 August 1987, p. 12.

[41]Horchler, p. 56.

[42]Wittig, p. 70; Ellen Feldman, *Collision Course: Plastic Packaging vs. Solid Waste Solutions* (New York: Environmental Action Coalition, 1989), p. 37.

[43]Letcher and Sheil, p. 227.

[44]Congressional Office of Technology Assessment, *Facing America's Trash,* Summary (Washington, D.C.: U.S. GPO, October 1989), p. 39.

[45]Miller, p. 499–501.

[46]Kirsten H. Oldenburg and Joel S. Hirschhorn, "Waste Reduction: A New Strategy to Avoid Pollution," *Environment,* 29, No. 2 (1987), 17–20 and 39–45.

[47]Ibid.

[48]U.S. EPA, "People and Places in the News: Dow, 3M, DuPont," *Pollution Prevention News* (Washington, D.C.: Pollution Prevention Office, U.S. EPA, May 1989).

[49]U.S. GAO, *Hazardous Waste, New Approach Needed to Manager the Resource Conservation and Recovery Act,* GAO/RCED-88-115, July 1988.

[50]U.S. EPA, "Waste Minimization Findings and Activities," Fact Sheet, October 1986.

[51]Joel Hirschhorn, "Cutting Production of Hazardous Waste," **Technology Review,** April 1988, pp. 52–61.

[52]U.S. EPA, **Report to Congress: Minimization of Hazardous Waste, Executive Summary,** (Washington, D.C.: U.S. EPA, October 1986).

[53]Malcolm S. Forbes, "Fact and Comment," **Forbes,** 10 August 1987, pp. 17–18.

[54]U.S. Office of Technology Assessment, **Technologies and Management Strategies for Hazardous Waste Disposal,** Summary (Washington, D.C.: U.S. GPO, March 1983).

[55]Ibid, p. 12.

[56]Ibid.

21. The Future: Altering the Course

[1]G. Tyler Miller, **Living in the Environment,** 3rd ed. (Belmont, CA: Wadsworth Publishing Company, 1982), pp. 485–486.

[2]G. Tyler Miller, **Living in the Environment,** 5th ed. (Belmont, CA: Wadsworth Publishing Co., 1988), p. 10.

[3]Lynn White Jr., "The Historical Roots of Our Ecological Crisis," **Science,** 155, 10 March 1967, 1203–1207.

[4]Miller, 1982, p. 60.

[5]Miller, 1988, p. 37.

[6]H.W.F. Saggs, **The Greatness That Was Babylon** (New York: Hawthorne Books, 1962).

[7]Lester Brown, and Sandra Postel, "Thresholds of Change," In: **State of the World 1987** (New York: W.W. Norton and Company, 1987), p. 16.

[8]Cynthia Pollock, "Realizing Recycling's Potential," In: **State of the World 1987** (New York: W.W. Norton and Company, 1987), p. 121.

[9]Sheryl Harris, "Reckless Misuse of Cheap Energy to Be Costly to All," **Akron Beacon Journal,** 10 December 1988, p. 5.

[10]**World Resources 1988–89** (New York: Basic Books, 1988).

[11]Ellen B. Geld, "Will Farming Destroy Brazil's Amazon Basin?" **The Wall Street Journal,** 13 January 1989, p. A13.

[12]Cynthia Pollock Shea, "Building a Market for Recyclables," **Worldwatch,** 1, No. 3 (1988), 15.

[13]Edward Flattau, "Loggers Could Save National Forests," **The Cleveland Plain Dealer,** 23 September 1988, p. 17A.

Bibliography

Am. Pub. Health Assoc., Am. Water Works Assoc., Water Pol. Control Federation, **Standard Methods for the Examination of Water and Wastewater,** 16th ed., Washington, DC: APHA, 1985.

Allen, M.J., and Geldreich, E.E., "Evaluating the Microbial Quality of Potable Waters," In: **Evaluation of the Microbiology Standards for Drinking Water,** C.W. Hendricks, ed., EPA-570/9-78-OOC, Washington, DC: U.S. EPA, 1978.

Brady, Patrick, "Waterborne Giardiasis," **Annals of Internal Medicine,** 81 (1974), 498–499.

Conservation Foundation, **Groundwater Protection,** Washington, D.C.: The Conservation Foundation, 1987.

Council on Environmental Quality, **Integrated Pest Management,** Washington, DC: U.S. GPO, December 1979.

Crump, Kenny S. and Guess, Harry A., "Drinking Water and Cancer: A Review of Recent Epidemiological Findings and Assessment of Risks," **Ann. Rev. Public Health,** 3, 1982, pp. 339–357.

Epstein, Samuel S., **The Politics of Cancer,** Garden City, NY: Anchor Books, Anchor Press/Doubleday, 1979.

Fiore, J.V., and Babineau, R.A., "Effect of an Activated Carbon Filter on the Microbial Quality of Water," **Appl. Env. Micro.,** Nov. 1977, pp. 541–546.

Hallaway, Joann, "Drinking Water Treatment Devices: Filters," No. 9.728, Colorado State University Extension Service, 1983.

Hen, J.D., and Rurum, W.H., "Solubility and Occurrence of Lead in Surface Water," **JAWWA,** August 1973, pp. 562–568.

Hoff, John C., "The Relationship of Turbidity to Disinfection of Potable Water," In: **Evaluation of the Microbiology Standards for Drinking Water,** Charles W. Hendrick, ed., U.S. EPA 570/9-78-OOC, Washington, D.C., 1978.

Horton, Thomas R., "Methods and Results of EPA's Study of Radon in Drinking Water," EPA 520/5-83-027, U.S. EPA, Office of Radiation Programs, Montgomery, AL, December 1983.

Jekel, Martin R., and Roberts, Paul V., "Total Organic Halogen as a Parameter for the Characterization of Reclaimed Waters: Measurement, Occurrence, Formation, and Removal," **Environ. Sci. & Technol.,** 14, No. 8 (1980), 970–975.

Keswick, Bruce H., and Gerba, Charles P., "Viruses in Groundwater," **Environ. Sci. & Technol.,** 14, No. 11 (1980).

Klancke, Douglas N., "Legionnaire's Disease: The Epidemiology of Two Outbreaks in Burlington, Vermont," **Am. J. Epidem.** 119, No. 3 (1980), 382–391.

Lappenbusch, William L., **Contaminated Drinking Water & Your Health,** Alexandria, VA: Lappenbusch Environmental Health, Inc., 1986.

Loper, John C., "Mutagenic Effects of Organic Compounds in Drinking Water," **Mutation Research,** 76 (1980), 241–268.

McDaniels, Audrey E., and Borduer, Robert H., "Effects of Holding Time and Temperature on Coliform Numbers in Drinking Water," **JAWWA,** September 1983.

U.S. EPA, **National Urban Pesticide Applicator Survey: Final Report, Overview & Results, Office of Pesticides and Toxic Substances,** Washington, DC: U.S. EPA, July 1985.

Patterson, James W., and O'Brien, Joseph E., "Control of Lead Corrosion," **JAWWA,** May 1979, pp. 264–271.

Miller, G. Tyler, **Living in the Environment,** 5th ed., Belmont, CA: Wadsworth Publishing Co., 1988.

National Academy of Sciences, **Drinking Water & Health,** Vol. 1, Safe Drinking Water Committee, National Research Council, Washington, DC: National Academy of Sciences, 1977.

National Academy of Sciences, *Drinking Water & Health,* Vol. 3, Safe Drinking Water Committee, National Research Council, Washington, DC: National Academy Press, 1980.

Regunathan, P., Beauman, W.H., and Kreusch, E.G., "Efficiency of Point-of-Use Treatment Devices," *JAWWA,* January 1983, pp. 42–50.

Reid, George W., and Lassovszky, Peter, "Treatment, Waste Management and Cost for Removal of Radioactivity from Drinking Water," *Health Physics,* 48, No. 5 (1985), 671–694.

Safe Drinking Water Committee, *Drinking Water & Health,* Volume 4, Washington, D.C.: National Academy Press, 1981.

Safe Drinking Water Committee, *Drinking Water & Health,* Volume 6, Washington, D.C.: National Academy Press, 1986.

Safe Drinking Water Committee, *Drinking Water & Health, Disinfectants & Disinfection Byproducts,* Volume 7, Washington, D.C.: National Academy Press, 1987.

Swallow, K.C., Shifrin, N.S., and Doherty, P.J., "Hazardous Organic Compound Analysis," *Environ. Sci. Technol.,* 22, No. 2 (1988), 136-142.

Taylor, Raymond H., Allen, Martin J., and Geldreich, Edwin, E., "Testing of Home Use Carbon Filters," *JAWWA,* 1979, pp. 577–579.

U.S. EPA, "Guidelines for Establishing Test Procedures for the Analysis of Pollutants Under the Clean Water Act; Final Rule and Interim Final Rule and Proposed Rule," *49 Federal Register,* 26 October 1984.

U.S. EPA, *Manual of Individual Water Supply Systems,* EPA/570/9-82-004, Washington, DC: Office of Drinking Water, October 1982.

U.S. EPA, "National Primary Drinking Water Regulations; Synthetic Organic Chemicals, Inorganic Chemicals and Microorganisms; Proposed Rule," *50 Federal Register,* 13 November 1985.

U.S. EPA, "National Primary and Secondary Drinking Water Regulations; Proposed Rule," *54 Federal Register,* 22 May 1989.

U.S. EPA, *Report to Congress: Waste Disposal Practices and Their Toxic Effects on Groundwater,* Report to Congress, Washington, DC: Office of Water Supply, January 1977.

U.S. EPA, *Seminar Publication: Protection of Public Water Supplies from Groundwater Contamination,* EPA/625/4-85/016, Washington, DC: U.S. EPA, September 1985.

U.S. Office of Technology Assessment, *Serious Reduction of Hazardous Waste: For Pollution Prevention and Industrial Efficiency,* OTA-ITE-317, Washington, D.C.: U.S. GPO, September 1986.

Appendix I

Part A
Additional Contaminants Required to Be Monitored By Some Water Companies

LIST 1	1,2-Dichloropropane	1,2-Dibromo-3-chloro-
Bromobenzene	1,3-Dichloropropane	propane (DBCP)
Bromodichloromethane	2,2-Dichloropropane	LIST 3
Bromoform	1,1-Dichloropropene	Bromochloromethane
Bromomethane	1,3-Dichloropropene	n-Butylbenzene
Chlorobenzene	Ethylbenzene	Dichlorodifluoromethane
Chlorodibromomethane	Styrene	Fluorotrichloromethane
Chloroethane	1,1,1,2-Tetrachloroethane	hexachlorobutadiene
Chloromethane	1,1,2,2-Tetrachloroethane	Isopropylbenzene
o-Chlorotoluene	Tetrachloroethylene	--Isopropyltoluene
p-Chlorotoluene	1,1,2-Trichloroethane	Naphthalene
Dibromomethane	1,2,3-Trichloropropane	n-Propylbenzene
m-Dichlorobenzene	Toluene	sec-Butylbenzene
o-Dichlorobenzene	p-Xylene	tert-Butylbenzene
trans-1,2-Dichloroethylene	o-Xylene	1,2,3-Trichlorobenzene
cis-1,2-Dichloroethylene	m-Xylene	1,2,4-Trichlorobenzene
Dichloromethane	LIST 2	1,2,4-Trimethylbenzene
1,1-Dichloroethane	Ethylene dibromide (EDB)	1,3,5-Trimethylbenzene

Part B
Additional Contaminants Required to Be Monitored By Some Water Companies

Priority I Contaminants	Sulfate	Diazinon	Malathion
ORGANICS	Thallium	Dichlofenthion	Merphos
Aldrin	Priority II	Dichloran	Methyl paraoxon
Butachlor	Contaminants	Dichlorvos	Methyl parathion
Carbaryl	PESTICIDES	Diphenamid	Mevinphos
2,4-DB	Ametryn	Diquat	MGK 264
Dalapon	Aspon	Disulfoton	MGK 326
Dicamba	Azinphos methyl	Disulfoton sulfone	Molinate
Dieldrin	BHC-alpha	Disulfoton	Napropamide
Dinoseb	BHC-beta	sulfoxide	Norflurazon
Hexachlorobenzene	BHC-delta	EPN	Pebulate
Glyphosphate	BHC-gamma	EPTC	cis-Permethrin
Hexachlorocyclopen-	Bolstar	Endosulfan I	trans-Permethrin
tadiene	Bromacil	Endosulfan II	Phorate
3-hydroxybarbofuran	Butylate	Endosulfan sulfate	Phosmet
Methomyl	Carboxin	Endothall	Prometon
Metribuzin	Chlorneb	Endrin aldehyde	Prometryn
Oxamyl (vydate)	Chlorobenzilate	Ethion	Pronamide
PAHs	Chloropropham	Ethoprop	Propazine
Phthalates	Chloropropylate	Ethylparathion	Simetryn
Picloram	Chlorothalonil	Etridiazole	Stirofos
Simazine	Chloropyrifos	Famphur	Tebuthiuron
2,3,7,8-TCDD	Coumophas	Fenamiphos	Terbacil
(Dioxin)	Cycloate	Fenarimol	Terbufos
2,4,5-T	DCPA	Fenitrothion	Terbutryn
INORGANICS	4,4'-DDD	Fensulfothion	Triademefon
Antimony	4,4'-DDE	Fenthion	Tricyclazole
Beryllium	4,4'-DDT	Fluridone	Trifuluralin
Cyanide	Demeton-O	Fonofos	Vernolate
Nickel	Demeton-S	Hexazinone	

Appendix II

SOC Monitoring Requirments

Bromobenzene	Dibromomethane	Ethylbenzene
Bromodichloromethane	m-Dichlorobenzene	Styrene
Bromoform	o-Dichlorobenzene	1,1,2-Trichloroethane
Bromomethane	trans-1,2-Dichloroethylene	1,1,1,2-Tetrachloroethane
Chlorobenzene	cis-1,2-Dichloroethylene	1,1,2,2,-Tetrachloroethane
Chlorodibromomethane	Dichloromethane	Tetrachloroethylene
Chloroethane	1,1-Dichloroethane	1,2,3-Trichloropropane
Chloroform	1,1-Dichloropropene	Toluene
Chloromethane	1,2-Dichloropropane	p-Xylene
o-Chlorotoluene	1,3-Dichloropropane	o-Xylene
p-Chlorotoluene	1,3-Dichloropropene	m-Xylene
	2,2-Dichloropropane	

Appendix III

SOC Tests Available

EPA 502.1 - Volatile Halogenated Organic Compounds

Chemical	DL (ppb)	CHemical	DL (ppb)
Bromobenzene	nd	1,1-Dichloroethane	0.002
Bromochloromethane	nd	1,2-Dichloroethane	0.002
Bromodichloromethane	0.002	1,1-Dichloroethene	0.003
Bromoform	0.02	cis-1,2-Dichloroethene	0.002
Bromomethane	0.03	trans-1,2-Dichloroethene	0.002
Carbon teetrachloride	0.003	1,2-Dichloropropane	nd
Chlorobenzene	0.001	1,3-Dichloropropane	nd
Chloroethane	0.008	2,2-Dichloropropane	nd
Chloroform	0.002	1,1-Dichloropropene	nd
Chloromethane	0.01	Methylene chloride	nd
2-Chlorotoluene	nd	1,1,1,2-Tetrachloroethane	nd
4-Chlorotoluene	nd	1,1,2,2-Tetrachloroethane	0.01
Dibromochloromethane	nd	Tetrachloroethene	0.001
1,2-Dibromoethane	0.03	1,1,1-Trichloroethane	0.003
Dibromomethane	nd	1,1,2-Trichloroethane	0.007
1,2-Dichlorobenzene	nd	Trichloroethene	0.001
1,3-Dichlorobenzene	nd	Trichlorofluoromethane	nd
1,4-Dichlorobenzene	nd	1,2,3-Trichloropropane	nd
Dichlorodifluoromethane	nd	Vinyl chloride	0.006

EPA 502.2 - Volatile Organic Components

Chemical	DL (ppb)	Chemical	DL (ppb
Benzene	0.009	1,2-Dichloropropane	0.006
Bromobenzene	0.03	1,3-Dichloropropane	0.03
Bromochloromethane	0.01	2,2-Dichloropropane	0.05
Bromodichloromethane	0.02	1,1-Dichloropropene	0.02
Bromoform	1.6	Ethylbenzene	0.005
Bromomethane	1.1	Hexachlorobutadiene	0.02
n-Butylbenzene	0.02	Isopropylbenzene	0.05
sec-Butylbenzene	0.02	p-Isopropyltoluene	0.01
tert-Butylbenzene	0.06	Methylene chloride	0.02
Carbon tetrachloride	0.01	Napththalene	0.06
Chlorobenzene	0.01	n-Propylbenzene	0.004
Chloroethane	0.1	Styrene	0.01
Chloroform	0.02	1,1,1,2-Tetrachloroethane	0.005
Chloromethane	0.03	1,1,2,2-Tetrachloroethane	0.01
2-Chlorotoluene	0.01	Tetrachloroethene	0.04
4-Chlorotoluene	0.01	Toluene	0.01
Dibromochloromethane	0.03	1,2,3-Trichlorobenzene	0.03
1,2-Dibromo-3-chloropropane	3.0	1,2,4-Trichlorobenzene	0.03
1,2-Dibromoethane	nd	1,1,1-Trichloroethane	0.03
Dibromomethane	2.2	1,1,2-Trichloroethane	nd

Source: U.S. EPA   nd - not determined

EPA 502.2 (cont'd)

1,2-Dichlorobenzene	0.02	Trichloroethene	0.01
1,3-Dichlorobenzene	0.02	1,2,3-Trichloropropane	0.4
1,4-Dichlorobenzene	0.01	1,2,4-Trimethylbenzene	0.05
Dichlorodifluoromethane	0.05	1,3,5-Trimethylbenzene	0.004
1,1-Dichloroethane	0.07	Vinyl chloride	0.04
1,2-Dichloroethane	0.8	o-Xylene	0.02
1,1-Dichloroethene	0.07	m-Xylene	0.01
cis-1,2-Dichloroethene	0.01	p-Xylene	0.01
trans-1,2-Dichloroethene	0.06		

EPA 503.1 - Volatile Aromatic and Unsaturated Organic Compounds

Chemical	DL (ppb)	Chemical	DL (ppb)
Benzene	0.02	4-Isopropyltoluene	0.009
Bromobenzene	0.002	Napthalene	0.04
n-Butylbenzene	0.02	n-Propylbenzene	0.009
sec-Butylbenzene	0.02	Styrene	0.008
tert-Butylbenzene	0.006	Tetrachloroethene	0.01
Chlorobenzene	0.004	Toluene	0.02
2-Chlorotoluene	0.008	1,2,3-Trichlorobenzene	0.03
4-Chlorotoluene	nd	1,2,4-Trichlorobenzene	0.03
1,2-Dichlorobenzene	0.02	Trichloroethene	0.01
1,3-Dichlorobenzene	0.006	1,2,4-Trimethylbenzene	0.006
1,4-Dichlorobenzene	0.006	1,3,5-Trimethylbenzene	0.003
Ethylbenzene	0.002	o-Xylene	0.004
Hexachlorobutadiene	0.02	m-Xylene	0.004
Isopropylbenzene	0.005	p-Xylene	0.002

EPA 524.1 - Volatile Organic Compounds (GC/MS)

Chemical	DL (ppb)	Chemical	DL (ppb)
Benzene	0.2	cis-1,2-Dichloroethene	nd
Bromobenzene	nd	trans-1,2-Dichloroethene	0.2
Bromochloromethane	nd	1,2-Dichloropropane	nd
Bromodichloromethane	0.2	1,3-Dichloropropane	nd
Bromoform	0.5	2,2-Dichloropropane	nd
Bromomethane	nd	1,1-Dichloropropene	nd
sec-Butylbenzene	nd	Ethylbenzene	nd
tert-Butylbenzene	nd	Hexachlorobutadiene	nd
Carbon tetrachloride	0.2	Isopropylbenzene	nd
Chlorobenzene	0.2	Methylene chloride	nd
Chloroethane	nd	n-Propylbenzene	nd
Chloroform	0.1	Styrene	nd
Chloromethane	nd	1,1,1,2-Tetrachloroethane	nd
2-Chlorotolune	nd	1,1,2,2-Tetrachloroethane	nd
4-Chlorotoluene	nd	Tetrachloroethene	0.2
Dibromochloromethane	0.2	Toluene	0.2
1,2-Dibromo-3-chloropropane	nd	1,1,1-Trichloroethane	0.2
1,2-Dibromoethane	nd	1,1,2-Trichloroethane	0.2
Dibromomethane	nd	Trichloroethene	0.2

EPA 524.1 (cont'd)

1,2-Dichlorobenzene	nd	Trichlorofluoromethane	nd
1,3-Dichlorobenzene	nd	1,2,3-Trichloropropane	nd
1,4-Dichlorobenzene	0.1	Vinyl chloride	nd
Dichlorodifluoromethane	nd	o-Xylene	nd
1,1-Dichloroethane	nd	m-Xylene	nd
1,2-Dichloroethane	0.2	p-Xylene	nd
1,1-Dichloroethene	nd		

EPA Method 601 - Purgeable Halocarbons

Chemical	DL (ppb)	Chemical	DL (ppb)
Chloromethane	0.08	1,1,1-Trichloroethane	0.03
Bromomethane	1.18	Carbon tetrachloride	0.12
Dibromochloromethane		Bromodichloromethane	0.10
Dichlorodifluoromethane	1.81	1,1,2-Trichloroethane	0.02
1,2-Dichloropropane		trans-1,3-Dichloropropéne	0.20
cis-1,3-Dichloropropane		2-Chloroethylvinyl ether	0.13
Vinyl chloride	0.18	Bromoform	0.20
Chloroethane	0.52	1,1,2,2-Tetrachloroethane	0.03
Methylene chloride	0.25	Tetrachloroethene	0.03
Trichlorofluoromethane	nd	Chlorobenzene	0.25
1,1-Dichloroethene	0.13	1,3-Dichlorobenzene	0.32
1,1-Dichloroethane	0.07	1,2-Dichlorobenzene	0.15
trans-1,2-Dichloroethene	0.10	1,4-Dichlorobenzene	0.24
Chloroform	0.05	Terichlorofluoromehtane	
1,2-Dichloroethane	0.03		

EPA Method 602 - Purgeable Aromatics

Chemical	DL (ppb)	Chemical	DL (ppb)
Benzene	0.2	1,4-Dichlorobenzene	0.3
Toluene	0.2	1,3-Dichlorobenzene	0.4
Ethylbenzene	0.2	1,2-Dichlorobenzene	0.4
Chlorobenzene	0.2		

EPA Method 604 - Phenols

Chemical	DL (ppb)	Chemical	DL (ppb)
2-Chlorohenol	0.31	4-Chloro-3-methylphenol	0.36
2-Nitrophenol	0.45	2,4-Dinitrophenol	13.0
Phenol	0.14	2-Methyl-4,6-dinitrophenol	16.0
2,4-Dimethylphenol	0.32	Pentachlorophenol	7.4
2,4-Dichlorophenol	0.39	4-Nitrophenol	2.8
2,4,6-Trichlorophenol	0.64		

EPA Method 606 (506) Phthalates

Chemical	DL (ppb)	Chemical	DL (ppb)
Dimethyl phthalate	0.29	Butyl benzyl phthalate	0.34
Diethyl phthalate	0.49	Bis (2-ethylhexyl) phthalate	2.0
Di-n-butyl phthalate	0.36	Di-n-octyl phthalate	3.0

EPA Method 608 - Organochlorine Pesticides and PCBs

Chemical	DL (ppb)	Chemical	DL (ppb)
a-BHC	0.003	4,4-DDT	0.012
'B-BHC	0.00	Endrin aldehyde	0.023
S-BHC	0.00	Endosulfan sulfate	0.066
Heptachlor	0.003	Chlordane	0.014
y-BHC	0.009	Toxaphene	0.24
Aldrin	0.004	PCB-1016	nd
Heptachlor epoxide	0.083	PCB-1221	nd
Endosulfan I	0.014	PCB-1232	nd
4,4-DDE	0.004	PCB-1242	0.065
Dieldrin	0.002	PCB-1248	nd
Endrin	0.006	PCB-1254	nd
4,4-DDD	0.011	PCB-1260	nd
Endosulfan II	0.004		

EPA 610 (550,550.1) - Polyaromatic Hydrocarbons (PAHs)

Chemical	DL (ppb)	Chemical	DL (ppb)
Naphthalene	1.8	Benzo(a)anthracene	0.013
Acenaphthylene	2.3	Chrysene	0.15
Acenaphthene	1.8	Benzo(b)fluoranthene	0.018
Fluorene	0.21	Benzo(k)fluoranthene	0.017
Phenanthrene	0.64	Benzo(a)pyrene	0.023
Anthracene	0.66	Dibenzo(a,h)anthracene	0.030
Fluoranthene	0.21	Benzo(ghi)perylene	0.076
Pyrene	0.27	Indeno(1,2,3-cd)pyrene	0.043

EPA 624 - Purgeables (GC/MS)

Chemical	DL (ppb)	Chemical	DL (ppb)
Chloromethane	nd	Trichloroethene	1.9
Bromomethane	nd	Benzene	4.4
Vinyl chloride	nd	Dibromochloromethane	3.1
Chloroethane	nd	1,1,2-Trichloroethane	5.0
Methylene chloride	2.8	trans-1,3-Dichloropropene	nd
Trichlorofluoromethane	nd	2-Chloroethylvinyl ether	nd
1,1-Dichloroethene	2.8	Bromoform	4.7
1,1-Dichloroethane	4.7	1,1,2,2-Tetrachloroethane	6.9
trans-1,2-Dichloroethene	1.6	Tetrachloroethene	4.1
Chloroform	1.6	Toluene	6.0
1,2-Dichloroethane	2.8	Chlorobenzene	6.0
1,1,1-Trichloroethane	3.8	Ethyl benzene	7.2
Carbon tetrachloride	2.8	1,3-Dichlorobenzene	nd
Bromodichloromethane	2.2	1,2-Dichlorobenzene	nd
1,2-Dichloroproane	6.0	1,4-Dichlorobenzene	nd
cis-1,3-Dichloropropene	5.0		

EPA 625 - Acid Extractables (GC/MS)

Chemical	DL (ppb)	Chemical	DL (ppb)
2-Chlorophenol	3.3	4-Chloro-3-methylphenol	3.0
2-Nitrophenol	3.6	2,4-Dinitrophenol	42
Phenol	1.5	2-Methyl-4,6-dinitrophenol	24
2,4-Dimethylphenol	2.7	Pentachlorophenol	3.6
2,4-Dichlorophenol	2.7	4-Nitrophenol	2.4
2,4,6-Trichlorophenol	2.7		

EPA 625 - Base/Neutral Extractables (GC/MS)

Chemical	DL (ppb)	Chemical	DL (ppb)
Acenaphthene	1.9	Benzo (a) anthracene	7.8
Dimethyl phthalate	1.6	3,3'-Dichlorobenzidine	16.5
2,6-Dinitrotoluene	1.9	Di-n-octyl phthalate	2.5
Fluorene	1.9	Benzo (b) fluoranthene	4.8
4-Chlorophenyl phenyl ether	4.2	Benzo (k) fluoranthene	2.5
2,4-Dinitrotoluene	5.7	Benzo (a) pyrene	2.5
Diethylphthalate	1.9	Indeno (1,2,3-c,d) pyrene	3.7
N-Nitrosodiphenylamine	1.9	Dibenzo (a,h) anthracene	2.5
Hexachlorobenzene	1.9	Benzo (ghi) perylene	4.1
b-BHC	—	N-Nitrosodimethylamine	—
4-Bromophenyl phenyl ether	1.9	Chlordane	—
S-BHC	—	Toxaphene	—
Phenanthrene	5.4	PCB-1016	—
Anthracene	1.9	PCB-1221	30
B-BHC	4.2	PCB-1232	—
Heptachlor	1.9	PCB-1242	—
S-BHC	3.1	PCB-1254	—
Aldrin	1.9	PCB-1260	36
Dibutyl phthalate	2.5	PCB-1260	—
Heptachlor epoxide	2.2	1,3-Dichlorobenzene	1.9
Endosulfan I	—	1,4-Dichlorobenzene	4.4
Fluoranthene	2.2	Hexachloroethane	1.6
Dieldrin	2.5	Bis (2-chloroethyl) ether	5.7
4,4'-DDE	5.6	1,2-Dichlorobenzene	1.9
Pyrene	1.9	Bis (2-chloroisopropyl) ether	5.7
Endrin	—	N-Nitrosodi-n-propylamine	—
Endosulfan II	—	Nitrobenzene	1.9
4,4'-DDD	2.8	Hexachlorobutadiene	0.9
Benzidine	44	1,2,4-Trichlorobenzene	1.9
4,4'-DDT	4.7	Isophorone	2.2
Endosulfan sulfate	5.6	Naphthalene	1.6
Endrin aldehyde	—	Bis (2-chloroethoxy) methane	5.3
Butyl benzyl phthalate	2.5	Hexachlorocylopentadiene	—
Bis (2-ethylhexyl) phthlate	2.5	2-Chloronaphthalene	1.9
Chrysene	2.5	Acenaphthylene	3.5

Appendix IV

Pesticide Methods

Method 504:

DBCP	EDB

Method 505:

Alachlor	Dieldrin	Hexachlorobenzene	PCBs
Aldrin	Endrin	Lindane	Simazine
Atrazine	Heptachlor	Methoxychlor	Toxaphene
Chlordane	Heptachlor Epoxide	Metolachlor	

Method 507:

Alachlor	Dichlofenthion	Fenthion	Pebulate
Ametryn	Dichlorvos	Fluridone	Phorate
Aspon	Diphenamid	Fonofos	Phosmet
Atraton	Disulfoton	Hexazinone	Prometon
Atrazine	Disulfoton sulfone	Malathion	Prometryn
Azinphos methyl	Disulfoton sulfoxide	Merphos	Pronamide
Bolstar	EPN	Methyl paraoxon	Propazine
Bromacil	EPTC	Methyl parathion	Simazine
Butachlor	Ethion	Metolachlor	Simetryn
Butylate	Ethoprop	Metribuzin	Stirofos
Carboxin	Ethyl parathion	Mevinphos	Tebuthiuron
Chloropropham	Famphur	MGK 264	Terbacil
Coumophos	Fenamiphos	MGK 326	Terbufos
Cycloate	Fenarimol	Molinate	Terbutryn
Demeton-O	Fenitrothion	Napropamide	Triademofon
Demeton-S	Fensulfothion	Norflurazon	Tricyclazole
Diazinon			Vernolate

Method 508:

Aldrin	DCPA	Endosulfan sulfate	Heptachlor
Chlordane-alpha	4,4-DDD	Endrin	Heptachlor epoxide
Chlordane-gamma	4,4-DDE	Endrin aldehyde	Hexachlorobenzene
Chlorneb	4,4-DDT	Etridiazole	Methoxychlor
Chlorobenzilate	Dichloran	BHC-alpha	cis-Permethrin
Chloropropylate	Dieldrin	BHC-beta	trans-Permethrin
Chlorothalonil	Endosulfan I	BHC-delta	Trifluralin
Chloropyrifos	Endosulfan II	BHC-gamma	

Method 515:

2,4-D	Dicamba	Dalapon	
2,4-DB	2,4,5-T	Pentachlorophenol	Dinoseb
	2,4,5-TP (Silvex)	(PCP)	Picloram

Method 531.1:

aldicarb	aldicarb sulfoxide	oxamyl (Vydate)	methomyl
aldicarb sulfone	carbofuran	carbaryl	

Appendix V

Labs That Specialize in Testing Water for Homeowners

Water Test
33 South Commercial Street
Manchester, NH 03101
1-800-426-8378

Suburban Water Testing Laboratories
4600 Kutzman Road
Temple, PA 19560
1-800-433-6595

National Testing Laboratories
6151 Wilson Mills Road
Cleveland, OH 44143
1-800-458-3330

Broward Testing Laboratories Inc.
P.O. Box 23541, 1034 N.E. 44th Court
Ft. Lauderdale, FL 33307
1-800-458-3330 / FAX (305) 776-0689

NOTE: *Inclusion on this list in no way implies an endorsement by the author or the publisher.*

Appendix VI

Companies Manufacturing Home Water Treatment Devices

Ametek, Inc.
(activated carbon, RO)
502 Indiana Avenue
P.O. Box 1047
Sheboygan, WI 53082-1047
(414) 475-9435, 1-800-222-7558

Atlantic Ultraviolet Corp.
(UV)
250 North Fehr Way
Bay Shore, New York 11706
(516) 586-5900

Bionaire Corp.
(activated carbon)
565 Commerce Street
Franklin Lakes, NJ 07417
(201) 337-2110, FAX (201) 337-9067

Coast Filtration
(RO, activated carbon, UV)
142 Viking Avenue
Brea, CA 92621
(714) 990-4602

Commercial Filters
(activated carbon)
State Road 32 West
Lebanon, IN 46052
(317) 482-3900

Creative Products Inc.
(distillation units)
3400 E. Century Avenue
Bismark, ND 58501
(701) 258-2621, 1-800-822-4229

Culligan International
(activated carbon, UV, RO)
One Culligan Parkway
Northbrook, Illinois 60062
(312) 498-2000

CUNO, Inc. Consumer Products
(RO)
400 Research Parkway
Meriden, Connecticut 06450
1-800-243-6894, In CT: (203) 237-5541

Durastill
(distillation units)
P.O. Box 76641
Atlanta, GA 30328
(816) 454-5260

Ecowater Systems
(activated carbon, RO, distillation)
P.O. Box 64420
St. Paul, Minnesota 55164
(612) 739-5330

Electrolux Water Systems Inc.
(UV, activated carbon)
2300 Windy Ridge Parkway
Suite 900
Marietta, Georgia 30067
(404) 933-1000

Environmental Purification Systems
(activated carbon, RO)
P.O. Box 191
Concord, CA 94522
(415) 682-7321, 1-800-829-2129

Everpure Inc.
(RO)
660 N. Blackhawk Drive
Westmont, Illinois 60559
(312) 654-4000

Filtration Marketing Services
(activated carbon, RO)
P.O. Box 9006
Lutherville, MD 21093-9789

General Ecology Inc.
(microfilters)
151 Sheree Boulevard
Lionville, Pennsylvania 19353
(215) 363-7900

Glacier Water Systems Inc.
(activated carbon)
550 Business Center Drive
The Kennsington Center
Mount Prospect, IL 60056
(708) 803-1150, FAX (708) 803-1186

Hague
(distallation units)
4343 South Hamilton Road
Graveport, OH 43125
(614) 836-2195, 1-800-282-3515

Hammacher Schlemmer
(distillation systems)
New York Store
147 East 57th Street
New York, NY 10022
(212) 421-9000, FAX (212) 644-3875

Hurley Water Systems
(activated carbon)
12621 South Laramie Avenue
Alsip, Illinois 60658-3225
(312) 388-9222

Ionics Incorporated
P.O. Box 99
Bridgeville, Pennsylvania 15017
(412) 343-1040, FAX (412) 257-1270

Kinetico Inc.
(activated carbon, RO)
10845 Kinsman Road
Newbury, OH 44065
(216) 564-9111, FAX (216) 564-9541

Multi-pure
(activated carbon, RO)
21339 Nordhoff
Chatsworth, CA 91311
(818) 341-7377, 1-800-622-9206

Nimbus Water Systems, Inc.
(RO)
P.O. Box 3263
Escondido, California 92025-0570
(619) 591-0211

Peridott
(RO, distillation unit in development)
3287 Kifer Road
Santa Clara, CA 95051
(408) 732-0612

Pure Water Inc.
(distillation systems)
3725 Touzalin Avenue, P.O. Box 83226
Lincoln, Nebraska 68501
1-800-842-5805

Rainsoft
(RO)
2080 E. Lunt Avenue
Elk Grove, Illinois 60007

Teledyne Water Pik
(activated carbon)
1730 East Prospect Street
Fort Collins, Colorado 80525-1310
(303) 484-1352

Ultra Dynamics Corp.
(UV)
299 W. Fort Lee Road
Bogata, New Jersey 07603
(201) 489-0044

Water and Health Inc.
(distillation units)
829 Lynn Haven Parkway, Suite 119
Virginia Beach VA 23452
1-800-222-7188, In VA: 1-800-523-6388

Waterwise Inc.
(distillation units)
26200 U.S. Highway 27
Leesburg, FL 34748
1-800-874-9028, FAX (904) 787-8123

Westbend Water Systems
(distillation units)
400 Washington Street
Westbend, WI 53095
(414) 334-6906

NOTE: *Inclusion on this list in no way implies an endorsement by the author or the publisher.*

Appendix VII
Sources of Recycled Products

Earth Care Paper Co.
(paper, paper products)
Write for thier free catalog:
100 South Baldwin Street
P.O. Box 3335, Dept. 462
Madison, WI 53704
(608) 256-5522

Seventh Generation
(assorted products)
Colchester, VT 05446-1672

Conservatree Paper Co.
(paper)
10 Lombard Street
Suite 250
San Francisco, CA 94111
1-800-522-9200

If you are having any printing done, whether for business or personal use, always request that the printer use recycled paper. Many conventional mills are now offering a variety of 50% to 100% recycled paper.

Appendix VIII

Environmental and Resource Preservation Organizations

The American Forestry Association
1516 P Street NW
Washington, DC 20005
Educates the public on sustainable use of forests and other natural resources.

Citizen's Clearinghouse for
   Hazardous Waste
P.O. Box 7097
Arlington, VA 22207
Provides information and assistance to grass roots organizations and individuals combatting irresponsible hazardous waste generation and disposal.

Clean Water Action Project
317 Pennsylvania Avenue SE, Suite 200
Washington, DC 20003
Lobbies for safer drinking water and laws to restrict pollution of waterways.

Defenders of Wildlife
1244 19th Street NW
Washington, DC 20036
Lobbies and educates the public on wildlife preservation.

Environmental Action Inc.
1525 New Hampshire Avenue NW
Washington, DC 20036
Research and education on a range of environmental issues.

Environmental Defense Fund Inc.
257 Park Avenue South
New York, NY 10010
Champions a wide range of environmental issues, through research, lobbying, and legal action.

The Friends of the Earth Environmental Policy Institute, Oceanic Society
218 D Street, SE
Washington, DC 20003
Lobbies, educates the public, and fights through legal means for protection of the oceans.

Greenpeace USA Inc.
1436 U Street NW
Washington, DC 20009
Involved in lobbying, environmental awareness, and education, and confrontation of polluters and hunters of endangered wildlife.

National Audobon Society
950 Third Avenue
New York, NY 10022
Public education, research, and lobbies Congress on many issues.

National Wildlife Federation
1400 16th Street NW
Washington, DC 20036-2266
Conducts research and educates the public on environmental issues.

Natural Resources Defense Council Inc.
40 West 20th Street
New York, NY 10011
Research, lobbying, and legal action on a number of environmental issues.

The Nature Conservancy
1815 N. Lynn Street
Suite 800
Arlington, VA 22209
Acquires threatened wilderness areas for protection.

Safe Water Foundation
6439 Taggart Road
Delaware, OH 43015
Concerned with safe drinking water, particularly the issue of fluoridation.

Sierra Club
730 Polk Street
San Francisco, CA 94109;
408 C Street NE
Washington, DC 20002
Lobbying, public education, and legal action on a number of environmental issues.

The Wilderness Society
1400 I Street NW, 10th Floor
Washington, DC 20005-2290
Lobbying, public education, and publishes reports mainly involving wilderness preservation.

Zero Population Growth
1400 16th Street NW, Suite 320
Washington, DC 20036
Concerned with population growth worldwide.

Glossary

activated carbon adsorption — The process of pollutants moving out of water and attaching on to activated carbon.

activated carbon — Very porous carbon from wood, coal, or lignite heated to very high temperatures to promote active sites where contaminants can be adsorbed from water.

active ingredient — The chemical in a pesticide formulation designed to kill a pest.

adsorption (adsorb) — Adhesion of chemicals out of water on to a surface.

aeration — The mixing or turbulent exposure of water to air and oxygen to dissipate volatile contaminants and other pollutants into the air.

aggressive water — Water which is soft and acidic and can corrode plumbing, piping, and appliances.

alkalinity — The measurement of constituents in a water supply which determine alkaline (opposite of acidic) conditions.

Alzheimer's Disease — A degenerative disease of the central nervous system characterized by premature senile mental deterioration.

aquifer — An underground waterway.

artesian well — A well drilled into a confined aquifer where enough pressure exists for the water to flow to the surface unaided.

backsiphonage — Reverse seepage of water in a distribution system.

backwashing — Reversing the flow of water through a home treatment device filter or membrane to clean and remove deposits.

bioaccumulation — The process by which certain metals and chemical levels in the tissue of organisms increase with higher standing in the food chain.

bioavailable — Chemicals or metals that are easily absorbed into the food chain or are easily taken up by the intestines in the human body.

biodegrade — To decompose by natural means.

biosphere — The earth and all its ecosystems.

blinds — Water samples containing a chemical of known concentration given a fictitious company name and slipped into the sample flow of the lab to test the impartiality of the lab staff.

brine — Highly salty and heavily mineralized water containing heavy metal and organic contaminants. Brine usually accompanies oil and gas deposits which exist far below drinkable groundwater supplies.

CERCLA — See Superfund.

CFCs (chlorofluorocarbons) — A class of compounds, used as refrigerants and in other chemical processes, which deplete the ozone layer in the stratosphere.

CDC — Centers for Disease Control.

CWSS — The national Community Water Supply Survey.

calcium carbonate — A white precipitate that forms in water lines, water heaters, and boilers, etc. in hard water areas; also called scale.

capillary zone — Soil area above the water table where water can rise up slightly through the cohesive force of capillary action.

carbamates — A class of new-age pesticides that attacks the nervous system of organisms.

carbonates — The collective term for the natural inorganic chemical compounds related to carbon dioxide that exist in natural waterways.

carcinogen — A substance that causes cancer.

cistern — A tank used to collect rainwater runoff from the roof of a house or building.

coagulation — In water treatment, the use of chemicals to make suspended solids gather or group together into small flocs.

cold vapor — Special method to test water for mercury.

coliform bacteria — Non-pathogenic microorganisms used in testing water to indicate the presence of pathogenic bacteria.

cone of depression — Natural depression in the water table around a well during pumping.

confined aquifer — An aquifer that lies between two impermeable rock layers.

confluent growth — In coliform testing, abundant or overflowing bacterial growth which makes accurate measurement difficult or impossible.

DNA — Deoxyribonucleic acid — basic genetic building block material in chromosomes of a cell nucleus.

daughter — The product formed by the decay of a radionuclide; usually a new element.

deionized water — Water free of inorganic chemicals.

dental fluorosis — Disorder caused by excessive absorption of fluorine and characterized by brown staining of teeth.

detection limit — The lowest level that can be determined by a specific analytical procedure or test method. The detection limit of a test is determined by the test method itself or the analytical instrument used in the test. The detection limit for TOX, for example, is generally considered to be around 5 ppb. Therefore, if no organics are detected the result will be expressed as < (less than) 5 ppb.

dispersion — The movement and spreading of contaminants out and down in an aquifer.

distillation — Water treatment method where water is boiled to steam and condensed in a separate reservoir. Contaminants with lower boiling points than water do not vaporize and remain in the boiling flask.

disinfection byproducts — Halogenated organic chemicals formed when water is disinfected.

distilled water — Water that has been treated by boiling and condensation to remove solids, inorganics, and some organic chemicals.

duplicates — Two separate samples with separate containers taken at the same time from the same place. This is a quality control method. The two results should be very close.

ecosphere — A plant, like the earth, which is capable of supporting life.

electrons — Negatively charged particles orbiting the nucleus of a molecule.

enteric viruses — A category of viruses related to human excreta found in waterways.

epidemiology — The science which investigates the origin of diseases or abnormalities affecting human populations.

fetotoxic — Toxic to the fetus.

flocculation — Large scale treatment process whereby small particles in flocs are collected into larger particles so their weight causes them to settle to the bottom of the treatment tank. This is accomplished by gentle stirring.

food chain — A biological hierarchy of organisms where each organism feeds on the one below it. (Example: grasses — rabbit — fox.)

fungicide — A pesticide specifically designed to kill fungus.

GAC — Granular Activated Carbon — pure carbon heated to promote "active" sites which can adsorb pollutants. GAC is used in some home water treatment devices to remove certain organic chemicals and radon.

GAO — General Accounting Office — Arm of Congress that investigates and evaluates government policy and programs.

gamma radiation — Electromagnetic ionizing radiation which easily penetrates biological tissue.

gastroenteritis — Infection and inflamation of stomach and intestines.

GI tract — Gastrointestinal tract — the stomach and intestines; the digestive tract.

geothermal — Related to heat from the interior of the earth.

greenhouse effect — Natural phenomenon whereby increased CO_2 and other gases in the atmosphere cause radiation from the sun to be trapped, leading to increases in global temperature and potential large-scale climate changes.

HPC — Heterotrophic Plate Count; see SPC.

half-life — The time it takes for one-half of a radioactive element or pesticide to decay.

halides or halogens — Chlorine, bromine, or fluorine.

halogenated organic chemical — Organic chemical containing chlorine, bromine, or fluorine.

hard water — Water containing a high level of calcium, magnesium, and other minerals. Hard water reduces the cleansing power of soap and produces scale in hot water lines and appliances.

heat of vaporization — The amount of heat necessary to convert a liquid (water) into vapor.

herbicide — A pesticide specifically designed to kill unwanted plants.

hexavalent chromium — Chromium (VI) — chromium that exists in the +6 state; hexavalent chromium is considered much more harmful than the other prevalent form, chromium (III).

hydrologic cycle — Natural pathway water follows as it changes between liquid, solid, and gaseous states.

IARC — International Agency for Research on Cancer.

impermeable — Not permeable; does not permit fluids to pass through.

indicator organisms — Microorganisms whose presence is indicative of pollution or of more harmful microorganisms. Coliforms are indicator organisms.

indicator tests — Test for a specific contaminant, group of contaminants, or constituent which signals the presence of something else (i.e., coliforms indicate the presence of pathogenic bacteria, high turbidity indicates the possible presence of organics and microbiological contamination, and a positive TOX result indicates the presence of manmade organic chemicals).

inert ingredient — Nonreactive components in a pesticide formulation or product used to "carry" the active ingredient.

inorganic chemicals — Those chemicals which do not contain carbon; these include nitrates, fluoride, and metals.

insecticide — A class of pesticides used to kill undesirable insects.

isotopes — Atoms of the same element which have the same number of protons but a different number of neutrons in the nuclei.

leachate — Water containing contaminants which leaks from a disposal site such as a landfill or dump.

MCL — Maximum Contaminant Level — the maximum level of a contaminant allowed in water by federal law. This level is based on health effects as well as currently available treatment methods.

mg/L — Milligrams of a contaminant per liter of water; same as parts per million (ppm.

manifest (hazardous waste manifest) — Written documentation of a hazardous waste shipment that must accompany the waste from generator, to transporter, to disposal facility, and be signed by representatives of each company.

methemoglobin — Compound formed by the oxidation of hemoglobin which reduces the ability of the blood to carry oxygen.

methemoglobinemia — Disease, usualy affecting infants which diminishes the oxygen carrying ability of the blood.

method blank — Laboratory grade water taken through the entire analytical procedure to determine if the samples are being accidently contaminated by chemicals in the lab.

mineralized tooth — A completely developed tooth.

mutagen — A chemical which causes a change in the genetic makeup of an organism which affects future generations.

mutagenic activity — The relative potential for a specific substance, material, or water sample to alter genetic structure and influence subsequent generations.

NAS — National Academy of Sciences.

NIPDWR — National Interim Primary Drinking Water Regulations.

NPDES *(permits)* — Issued under the National Pollutant Discharge Elimination System for those companies discharging pollutants directly into waterways in the United States.

neutrons — Uncharged particles found in the nucleus of an atom which contribute to the total weight of the atom.

nitrate — (NO_3) — A form of nitrogen most often found in water.

OTA — The U.S. Congressional Office of Technology Assessment. Arm of the

Congress that studies the beneficial and detrimental impacts of technological applications.

oncogenic — Used to describe a substance or chemical that causes tumors.

organic chemials — Defined as chemicals containing carbon. These include: 1.) natural — those from animal and plant life, including coal and oil; and 2.) those synthesized by man — industrial solvents, pesticides, etc.

organic farming — Farming technique which utilizes only natural means to fertilize soil and control pests.

organophosphates — A class of organic pesticides containing phosphorus, which interrupts nerve impulses along the central nervous system leading to convulsions, paralysis, and death.

ozone — 1.) In the lower atmosphere, a colorless toxic gas formed by the reaction of nitrogen oxides and hydrocarbons with sunlight, particularly in urban areas. It contributes to chronic human respiratory disease. 2.) In the stratosphere, absorbs ultraviolet radiation and protects life from excessive ultraviolet doses. 3.) Used as an alternate form of water disinfection.

PAHs (Polyaromatic hydrocarbons) — A class of chemical pollutants formed through the breakdown of other chemicals or substances. For example, the combustion of coal, wood, oil, and certain SOCs results in the formation of PAHs.

pH — A measure of the acidity or alkalinity of water; the pH scale is from 0 to 14 — seven is neutral, 0 is very acidic, 14 is very alkaline.

ppb — Parts per billion.

ppm — Parts per million.

parameter — Test or analytical procedure which: 1.) measures specific chemicals such as pesticides, PCBs or fluoride; or 2.) measures a group of chemicals or a characteristic such as pH, TOX, or hardness.

parent — In the process of radioactive decay of one element, a new one is formed. The original elemental is called the parent.

Parkinson's disease — A degenerative disease which causes deterioration of the brain and affects the muscles.

pathogen — Microorganism which can cause disease.

phenols — A group of SOCs used in the synthesis of a wide variety of chemicals and a ubiquitous class of pollutants.

phthalates — A group of SOC chemicals used in plastics manufacturing to make them flexible and pliable. One of the most widespread classes of pollutants due to their usage.

plug — Concentrated amount of a contaminant or contaminants existing in soil or groundwater.

plume — The area taken up by contaminant(s) in an aquifer.

point of entry (POE)— Water treatment device situated at the point where the water enters the house; treats all water entering the house.

point of use (POU) — Home water treatment devices located at the point where the water is used, at or near the faucet.

precipitate — A solid which has come out of an aqueous solution. For example, iron from groundwater precipitates to a rust colored solid when exposed to air.

preservative — A chemical added to a water sample to keep it stable and prevent compounds in it from changing to other forms or microorganism densities from changing prior to analysis.

protons — Positively charged particles found in the nucleus of an atom.

punitive damages — Monetary award to a plaintiff from the defendent by the court as a punishment for wrongdoing.

purge — To force a gas through the water sample to liberate volatile chemicals or other gases from the water so their level can be measured.

purgeable organics — Volatile organic chemicals which can be forced out of the water sample with relative ease through purging.

quadruplicate testing — Testing the same sample four times; this is often done as a quality control measure in screening tests like TOC and TOX.

quality assurance — Efforts by a laboratory to ensure their test results can be substantiated by other laboratories.

quality control — Actions taken by a lab to ensure all variables and factors are considered in the measurement of a sample and the interpretation of data to give a result.

quantify — To measure the amount of a chemical or substance in a sample.

RCRA — Resource Conservation and Recovery Act — federal legislation requiring that hazardous waste be tracked from "cradle" (generation) to "grave" (disposal).

radionuclides — Radioactive elements or atoms.

recharge zone — The area over which an aquifer is replenished; for a confined aquifer the area would be small, but for an unconfined aquifer the recharge zone would likely be the entire length of the aquifer.

renal tubular dysfunction — Relating to damage to the kidneys.

replicate (sample analysis) — Analyzing the same sample twice; should yield very similar results.

residual chlorine — The level of chlorine existing in the distribution system after chlorination at the drinking water treatment plant. The residual chlorine level will be a function of the level of microorganisms and the potential for additional THM formation in the distribution system.

reverse osmosis (RO) — A water treatment method whereby water is forced through a semipermeable membrane which filters out impurities.

SOCs — Synthetic Organic Chemicals — Manmade chemicals containing carbon, many are associated with chronic health effects.

SPC — Standard Plate Count or Heterotrophic Plate Count (HPC) — A test which directly measures the level of certain bacteria in a water sample.

screening test — A test that encompasses a wide range of possible contaminants. Example are: TOX, TOC, and VOA. See also indicator tests.

secondary pest — An organism (usually an insect) that has become a pest due to the use of chemical pesticides on another pest.

sedimentation — A large scale water treatment process where heavy solids settle out to the bottom of the treatment tank after flocculation.

skeletal fluorosis — A health effect of excess fluoride leading to rheumatic effects, pain, and stiffness.

specific heat — The amount of heat required to raise the temperature of a kilogram of a substance (water) by 1°C.

spikes — A quality control measure where a known amount of chemical is added to the sample to determine how well the chemical is recovered from the sample when analyzed.

Superfund — Comprehensive Environmental Response, Compensation and Liability Act (CERCLA) — gave EPA authority and money to clean up abandoned, leaky hazardous waste sites.

surface impoundment — An indented area in the land's surface — pit, pond, lagoon — usually unlined and confined by natural means and holds liquid waste.

THMs — Trihalomethanes — a class of volatile organic chemicals created as a result of water chlorination.

TIC — Tentatively Identified Compounds — In GC/MS analysis, chemicals identified by computer match that are not covered in the specific test method. The accuracy of TIC levels is questionable.

TNTC — Too numerous to count — a total coliform test result; too many coliforms to count indicates heavy contamination.

TOC — Total Organic Carbon — screening test which measures the amount of organic carbon in the water sample.

TOX — Total Organic Halide — screening test which measures the level of organic chemicals containing chlorine and bromine.

teratogen — A chemical or substance which causes abnormal formation of a fetus.

total dissolved solids (TDS) — The sum of all inorganic and organic particulate material; TDS is an indicator test usually reserved for wastewater analysis, but is also a measure of the mineral content of bottled water and groundwater.

toxic tort — Wrongful injury or damage involving toxic chemicals leading to a civil law suit.

USGS — United States Geological Survey.

UV — Water disinfection treatment method using ultraviolet light.

unconfined aquifer — An aquifer that is not confined by impermeable rock above it so water recharge occurs across its entire length.

VOA (*Volatile Organic Analysis*) — Testing procedure for volatile organic chemicals; also called volatiles scan, volatiles screen, or referred to by specific EPA method number, EPA 601 or EPA 602.

volatile organic chemical (VOC) — An organic chemical which can easily dissipate or evaporate into the air.

water filter — Broad term used to describe different types of water filters. GAC filters are among the most widely used filters to treat drinking water.

water table — The surface of an unconfined aquifer which fluctuates due to seasonal precipitation.

wet methods — A group of water tests usually for determining the presence of inorganic chemicals, like nitrates, fluoride, and dissolved solids. They are called wet methods presumably due to the fact that they generally do not require the use of solvent extraction methods that SOC testing does.

Index

About the Author . . .

John Cary Stewart is an Environmental Chemist with 10 years of laboratory testing and consulting experience. He holds B.S. degrees in Chemistry and Water Analysis from California State University of Pennsylvania and has completed graduate work in Environmental Chemistry. He lives in Garrettsville, Ohio with his wife Sharon.

ORDER FORM

Please send me *Drinking Water Hazards*, by John Cary Stewart. Enclosed is $14.95 per copy plus $1.50 shipping. Please allow 4–8 weeks for delivery. Send check or money order payable to Envirographics.

number of copies _____

x 14.95

subtotal _____

shipping _____

Ohio residents please add 6.5% sales tax _____

Total amount enclosed _____

SHIPPING ADDRESS:
(please print clearly)

name _____

address _____

city / state _____

zip code _____

Mail to:
Envirographics • P.O. Box 334 • Hiram, Ohio 44234